Smart Yogurt

New Ways to Make Yogurt That Minimize Prep,
Optimize Output, Improve Taste and Texture,
Add Natural Flavors, Reduce Intolerance,
and Boost Probiotics

— Mark Shepard —

Yogurt is the quickest and simplest of all the fermented foods you can make at home—or it *should* be. More recent methods tend to complicate the process or make it less efficient. *Smart Yogurt* strips away unnecessary steps and identifies the most practical equipment, so your prep takes no more than a few minutes.

But maybe you have special needs or goals for your yogurt. For those who want to go beyond basics, *Smart Yogurt* presents a wealth of possibilities:

- Improving taste and texture without adding to prep time.
- Adding natural flavors and colors before incubation, without interfering with firming.
- Making your own lactose-free yogurt to increase tolerance.
- Boosting your yogurt's probiotic value by starting it from scratch—without any yogurt starter, dried "heirloom" culture, or probiotic tablets.
- Making your own non-dairy yogurt without thickeners or stabilizers.

Whether you want to simplify your yogurt making, explore new options, or just understand all the ways milk can be turned into one of the world's most popular foods, *Smart Yogurt* is your guide.

"Slim but thorough . . . This valuable guide's explanations and examples will inspire both new and veteran yogurt makers."

Kirkus Reviews

Books by Mark Shepard

Cookbooks
Smart Yogurt
Smart Sourdough
Simple Sourdough

Music
How to Love Your Flute
Simple Flutes

Alternatives
Gandhi Today
The Community of the Ark
Mahatma Gandhi and His Myths

Poetry
Songs of Flesh, Songs of Spirit

Smart Yogurt

New Ways to Make Yogurt that
Minimize Prep, Optimize Output,
Improve Taste and Texture,
Add Natural Flavors,
Reduce Intolerance, and
Boost Probiotics

Mark Shepard

Shepard Publications
Bellingham, Washington

For updates and more resources,
visit Mark's Sour Foods Page at

www.markshep.com/sour

Copyright © 2025 by Mark Shepard
Permission is granted to copy or reprint portions for any noncommercial use, except they may not be posted online without permission.

Library of Congress Control Number: 2024952580
Library of Congress subject headings: Yogurt; Cooking (Yogurt); Fermented milk; Fermented foods; Fermentation

Version 1.2

Acknowledgments

Thanks for inspiration go to Harold McGee, Sandor Ellix Katz, and David Asher. Thanks for valuable feedback and support go to Holly Howe, and above all, to my beloved wife, Anne L. Watson.

Cover Photos

Front top: Smart Mediterranean Yogurt, tilted
Front bottom: Warming devices
Back: Yogurt "rainbow" with natural colors

Contents

Getting Started	5
What Is Yogurt?	7
Yogurt Facts & Fictions	8
Minimizing Prep	14
Smart Yogurt	*18*
Optimizing Output	23
The Milk Carton Gambit	27
Optimum Temperature	33
Improving Taste & Texture	35
The Low-Down on Ultra-Filtered	39
Smart Greek Yogurt	*42*
Smart French Yogurt	*45*
Smart Mediterranean Yogurt	*46*
Adding Natural Flavors	47
Dissolving a Powder	51
Juice—A Buyer's Guide	54
Adding Natural Colors	63

Reducing Intolerance	65
Smart Lactose-Free Yogurt #1	*67*
Smart Lactose-Free Yogurt #2	*68*
Friend or Foe?	71
Boosting Probiotics	72
What About Raw Milk?	76
Strength or Diversity?	92
APPENDIX: The Non-Dairy Option	93
Smart Soy Milk Yogurt	*95*
Index	96

Getting Started

Let's handle a few basic questions up front.

Why eat yogurt? It's a tasty food that's both healthful and satisfying. It has all the nutrients of regular milk and more, in a form that may be digestible even for those who cannot otherwise tolerate dairy. Most yogurt also contains live bacteria that can be beneficial to gut health.

Why make your own? One reason might be saving money. The cost of homemade yogurt could be as low as a quarter of the cost of store-bought. And it's never likely to rise above half, even if you splurge on premium milk.

Another reason might be to avoid commonly added ingredients like thickeners and stabilizers. You might also want a specialized yogurt that's hard or impossible to get commercially—a yogurt with even richer taste, or with lactose more heavily reduced, or with greatly strengthened probiotics.

Why not *make your own?* There are good reasons here too. Most home recipes are exacting and time-consuming. And in the end, you may wind up with a product you like *less* than commercial yogurt, with little or no added health benefit.

I stopped making my own yogurt for years, because it didn't seem worth the effort. Eventually, though, my experiments convinced me otherwise—and led me to write this book. Here I hope to put to rest your reasons *against* and strengthen your reasons *for*, while also exploring possibilities that might be entirely new to you.

I'll start by showing how to cut your total prep and processing time to about five minutes, with no thermometer required.

Next, I'll present several modern devices that can give you more control over the process and allow much larger batches.

Then you'll find out how to thicken your yogurt without straining it or adding any thickener or stabilizer—not to mention the extremely simple way to make it tastier and more satisfying than most yogurt you can buy. Plus how to add your choice of natural flavors at the start and still get firm yogurt.

After that, I'll tell how to make your yogurt even more easily digestible, for those who have trouble with dairy. Then I'll present several new methods of my own for increasing the diversity of your yogurt's beneficial bacteria. These methods will provide probiotic value far beyond that of *any* yogurt in a store—or for that matter, any homemade yogurt touted by today's health gurus and influencers.

Finally, for those looking for vegan alternatives, I'll explore the possibility of making a true yogurt—not just an imitation!—that's dairy free and plant based.

Yogurt is likely already a rewarding part of your diet and lifestyle. Let's make it even more so.

What Is Yogurt?

Already know what yogurt is? Or *think* you know? Let's find out . . .

Milk proteins—like proteins in general—have molecules with tight structures, keeping them self contained and suspended in the milk. But like proteins in general, they can be *denatured*—a process commonly described as unfolding, unwinding, unraveling, unfurling. Basically, the molecules loosen up and stretch out, exposing parts of themselves that were out of reach. Those parts can then attach to one another so the proteins link up.

Different proteins get denatured in different ways. For casein—the main milk protein—it's mostly by exposure to acid. Of course, that's exactly what you get in yogurt making. As bacteria ferment the milk, they consume the milk sugars and produce lactic acid as a waste product. Eventually, there's enough acid to denature the casein, causing its molecules to link in a mesh.

And there you have your yogurt. While lactic acid provides the distinctive sourness, a mesh of linked casein provides the firmness.

Much as a sponge holds water, this protein mesh holds the remaining parts of the milk, namely the milk fat and the whey. *Whey* is itself a mixture of water and anything from the milk that can still dissolve in it. That includes sugars, minerals, the proteins that *didn't* denature, and all the lactic acid.

As you probably know, whey commonly leaks from the yogurt and makes puddles, and it's often partially drained off to thicken the yogurt. But don't think of whey as a waste product, as it would be if you were making cheese. It's a part of the yogurt itself—and a highly nutritious one.

Yogurt Facts & Fictions

Yogurt, like any health food or any fermented food made at home, has collected a lot of myths around it. Let's try to replace those myths with facts and common sense.

Myth: Greek yogurt is yogurt that's strained.

Almost any recipe for homemade Greek yogurt will tell you this. But most store-bought brands of Greek yogurt are *not* strained. And in Greece itself, strained yogurt is only one way traditional yogurt was made—and not the most common one.

So, it's best to think of Greek yogurt simply as yogurt that's thicker than most. There are several ways to achieve this, including a couple of ways much simpler than straining.

Myth: French yogurt is creamy because it's made in the cup.

Isn't most homemade yogurt made in small containers? And does that make it creamy? No. The French yogurt you can buy is simply yogurt that's thicker and creamier than most. And just as with Greek yogurt, this can be achieved in several ways.

In fact, commercial French and Greek yogurts are often made with the same ingredients and methods. They could just as well be labeled either way!

Myth: For yogurt to be yogurt, it must contain two particular bacteria species.

When yogurt was first developed as a commercial dairy product, it was made with just two bacteria species isolated from the many found in traditional yogurt. So, when U.S. government regulators created a standard definition for this product, they based their definition on those two species. As a result, any product today sold in the U.S. as "yogurt" must by law include those two, no matter what other species it has.

But this has nothing to do with yogurt as a traditional food. As long as you add enough lactic acid bacteria of *any* species to regular milk, it should sour and firm up as some sort of yogurt.

Myth: If you use store-bought yogurt as your starter, you can't keep taking yogurt from each batch to start the next.

Starting originally with commercial yogurt, neither my wife nor I have ever found a limit to the number of times we can use yogurt from one batch to start another. Contrary to claims, you absolutely do *not* need an "heirloom" yogurt to do this.

If your culture loses strength, the culprit may be temperature. The bacteria species in commercial yogurt are selected to be most active at temperatures over 100°F, so incubating at lower ones may gradually weaken your culture. And as you'd expect, temperatures too high can do the same.

Though this is strictly optional, you can ensure consistency by starting each time with your favorite commercial yogurt. You don't need more than a tablespoon of it to make a large jar of your own, and the opened tub should keep for at least a month in the refrigerator.

Myth: You can cut incubation time in half by doubling the amount of yogurt you use as starter.

Not nearly! Incubation time is less affected by starter amount than by fermentation speed, which is based on the bacteria's *doubling time*—the time it takes for the bacteria population to double. That time varies by species and by temperature, but as an overall ballpark figure for yogurt making, I figure it at about an hour.

So, let's say you add one tablespoon of yogurt to milk and quickly warm this culture to a typical temperature. After one hour, the bacteria population will have about doubled, giving you the equivalent of *two* tablespoons of starter. Say you now mix a second culture but this time add two tablespoons to start it. The remaining incubation time of both cultures will be the same!

In other words, doubling the starter will save you only about an hour of incubation.

Myth: Milk must be heated high and then cooled back down before adding the yogurt starter.

Many people believe that this preheating, or *scalding*, is needed to kill off competing microbes. But assuming your milk is pasteurized and reasonably fresh, nothing in it is going to catch up with the bacteria population you're adding with your starter. Those bacteria themselves will discourage anything else from settling in.

There's another purpose to scalding, though: It alters some of the milk protein so there's more of it to help firm up the yogurt. But this is only one way to increase firmness, and it's not the most efficient. Frankly, I don't advise even considering it unless you're making low-fat or nonfat yogurt, which tend to be runny. Otherwise, it's not worth the trouble.

Instructions for *some* electric yogurt makers tell you to scald the milk, while instructions for others omit that. This does *not* mean that one yogurt maker needs a different method than another. The difference is just in the instructions, not in the yogurt makers! You can always choose not to scald.

Myth: If your temperature is more than a few degrees off, your milk won't ferment.

The heat-loving bacteria in commercial yogurt will stay active in anything from room temperature to around 130°F (or 55°C), where they may start to weaken or die off. Within that range, the question isn't *if* the bacteria will ferment your milk, but how fast.

Every bacteria species has an optimum temperature, and the farther you deviate from that—higher or lower—the more you slow it down. But the difference has to be extreme before activity stops entirely.

By the way, the common warning that temperatures above 120°F (or 45°C) will kill your bacteria is *not* valid for the heat-loving bacteria in commercial yogurt. Some might even survive past 140°F (60°C)—though I haven't felt a need to test that.

Myth: Yogurt bacteria feed on lactose, so you can't make yogurt from lactose-free milk.

While *some* yogurt bacteria feed on *lactose*—milk sugar—they can *all* feed on less complex forms of sugar. At the same time, some or all of the lactose in lactose-free milk isn't actually removed—it's just broken down into some of these simpler forms.

So, lactose-free milk always has *some* kind of sugar the bacteria can feed on. If it didn't, it would be too bland to drink!

Don't be confused by the name *lactic acid bacteria*. Lactic acid is what these bacteria produce as waste, and they don't need lactose to do it.

Myth: Yogurt making needs bacteria that ferment milk and milk alone.

If I told you how many decades I believed this, I'd be giving away my age. But the truth is, yogurt is made with lactic acid bacteria of the very same species found in naturally fermented pickles, sauerkraut, and sourdough starter—though it's usually made today with far *fewer* species than are in those other foods.

Myth: Yogurt is probiotic, so it's good for gut health.

This is true to a point, of course, but not as much as believed. Yes, commercial yogurt with live cultures—as well as any homemade yogurt started from it—will contain around two to six bacteria species that are good for your gut. But traditional yogurt—just like naturally fermented sauerkraut, pickles, and the like—will have some or all of those species plus dozens more!

So, the yogurt you buy, and the yogurt you start from it, does not rank that high on the probiotics scale. And ironically, the same goes for yogurt made from probiotics supplements, which might provide only marginally more species, or only a single one!

Now, if you're eating a variety of naturally fermented foods, or even just raw, whole fruits and vegetables, you might not need your yogurt to supply a wide range of bacteria. But in case you do, I'll tell you how to easily make your own full-spectrum probiotic yogurt from scratch—with no need of an "heirloom" yogurt as starter.

Myth: To boost its probiotic value, yogurt should ferment for 24 hours, or 36.

If you ferment milk with lactic acid bacteria for long periods at anywhere near typical incubation temperatures, you will get massive population growth *at first*. But this will be followed by leveling off and then massive die-off, as the population outstrips its food supply or is finally overwhelmed by the lactic acid it's producing as waste.

Advocates of long ferments generally back up their claims with theoretical calculations based on growth rates in ideal conditions—not in the real world, where growth is never limitless! They may also cite bacteria counts from lab tests, not mentioning that those tests count dead bacteria along with the living. And their recipes often include a stabilizer like inulin, which makes sure that later batches firm up even if no bacteria survived the first.

For greatest probiotic strength, a culture should be warmed only as long as the bacteria are still multiplying faster than they're dying—then chilled before that's reversed. Warming for 24 hours or more is a recipe for crippling a culture.

Myth: Any animal or plant milk can be made into yogurt.

To make true yogurt, the milk must have proteins that react to the acid produced by bacteria, causing the proteins to link up in a firm mesh. Animal milks have such proteins, but most plant milks do not. Most so-called dairy-free yogurts, even if soured by bacteria, are actually puddings firmed up with thickeners or stabilizers.

For that reason, this book will focus on yogurt made from regular dairy milk—though at the end, I'll share one recipe for honest-to-goodness vegan yogurt.

Myth: Saturated fat is bad for you, so you should stick to low-fat or nonfat dairy—including yogurt.

That has been standard dietary guidance since the 1980s. But the recommendation was based on studies of saturated fat in general, not in specific whole foods, which may in some way change or balance its effects. (In other words, saturated fat in meat could be a whole lot worse for you than saturated fat in milk or yogurt.)

Research since then on dairy products has failed to show that full-fat dairy is bad for your health, and in fact has suggested the opposite. Many nutritionists now call for narrowing the original advice or discarding it altogether.

Obviously, the extra calories in full-fat dairy can lead to weight gain. But people who save calories from fat will often make it up in refined starches and sugars, which can be worse. So, avoiding full-fat yogurt may only deprive you of the tastiest, most satisfying, most natural, and even healthiest form of yogurt.

Minimizing Prep

Heat the milk high, but not too high. Cool it back down, but not too far. Stir during heating and cooling to keep skin from forming, or remove it later, or both. After setting, strain the result and deal with the whey.

When did yogurt making get so *fiddly*?

Answer: When it became a diet food.

Reclaiming Yogurt

You see, when yogurt first spread beyond its traditional home countries, it was not sold as a diet food at all. It was sold as a *health* food. To later make it a *diet* food, it had to be made from low-fat or nonfat milk. And there was just one problem: Straight low-fat or nonfat yogurt is not very good.

So now, commercial and home yogurt makers had to jump through hoops just to make their yogurt palatable. And that's how you got these unnecessary and time-consuming steps added to what should be one of the simplest, quickest kitchen procedures around.

How do we reclaim the ease of earlier methods? Two words: Whole milk.

By accepting a normal amount of fat content in our yogurt, we can strip down the process till it's once again quick and easy. Our yogurt will be thicker, taste better, satisfy longer, and truthfully, probably won't have that many more calories than nonfat yogurt with its added milk powder, fruit, or straining—all meant to compensate for what's now lacking.

So, even though this book is devoted to *new* approaches to making yogurt, our starting point is to reclaim an *old* method. Here, then, using whole milk, is the simple, quick, and easy way yogurt *used* to be made at home.

The Basics

Here are the basics you'll need to know before starting this recipe. Though the initial setup might seem intimidating, the prep time for later batches should only be about five minutes per jar!

Ingredients

You don't need to get fancy with the ingredients, at least for now.

When I say to use whole milk, I mean any fresh pasteurized, homogenized cow's milk with normal fat content—in other words, the regular kind of whole milk you buy in a store, whether from an independent brand or the store's private label.

Please don't use low-fat or nonfat or powdered milk. You should also avoid UHT (Ultra High Temperature) or sterilized milk—not because you can't make yogurt from it, but because nutrients are lost in the extreme heating. Ultra-filtered milk also has drawbacks, as I'll discuss later.

Raw milk, regardless of health benefits, has microbes and enzymes that interfere with yogurt making—so, this milk won't work unless you first pasteurize it yourself. And non-dairy milks won't work at all for *real* yogurt—with one exception, again discussed later.

"Yogurt with live cultures" means any yogurt that has not been pasteurized *after* the yogurt was made. To be clear, nearly all yogurt is made from pasteurized milk, but that's milk pasteurized *beforehand*. If the yogurt was also pasteurized afterward, then most of the yogurt bacteria will be dead, and it will be useless for making your own.

If the yogurt contains live cultures, the label will usually say so. If you're not sure, don't use it! Of course, if your yogurt comes from a batch you made yourself, you're safe. (That is, unless you've taken misguided advice, like instructions to ferment your yogurt for 24 hours or more.)

My recipe calls for plain yogurt as starter, and that is normally what you'd use. But since you need so little, it won't hurt if your starter comes with a little sweetening or flavoring. For instance, I've sometimes used honey yogurt or vanilla yogurt, or even taken yogurt from the top of a cup that had fruit at the bottom.

Equipment

You'll need some way to incubate the yogurt, keeping it warm while it's fermenting. If you've already been making yogurt, then whatever equipment or method you've used for that should be fine. In that case, you can adjust the recipe amounts to fit your usual container.

For those who don't have such equipment, I'll discuss options in the next chapter. Meanwhile, my recipe assumes you have a non-reactive container that holds a bit more than three cups of liquid. My standard solution for this is a wide mouth quart Mason jar with a plastic screw-on lid.

Food Safety

Though my recipe does *not* start with heating the milk to a high temperature, you can add that step if you're worried about the freshness of your milk—or as I said, if you're using raw milk. A temperature of 160°F or 70°C will pasteurize the milk in less than a minute. Just make sure you let the milk cool back down to a normal incubation temperature before adding your starter. (You might also have to discard "skin" that forms on top.)

Generally, your yogurt container only has to be washed normally—by dishwasher or by hand—to be safe for fermenting. If you have any doubt about it, it's probably enough to just fill the container with your hottest tap water and let it sit for a couple of minutes. To be even safer, you could heat the container to 160°F or 70°C, just as with milk.

With the temperatures used for incubating, it's understandable to be concerned about food safety. But for yogurt, your milk and your equipment do *not* need to be 100% sterile, or even close. That's because the yogurt bacteria themselves discourage other microbes.

Keep in mind, yogurt didn't start out as a way to create a tasty, healthful dish. It started out as a way to preserve milk! So, to a large extent, you can rely on the yogurt bacteria themselves to keep your yogurt safe.

Smart Yogurt

    3 cups pasteurized whole milk
    1 tablespoon plain yogurt with live cultures

1. Pour the milk into your yogurt container. The amount does *not* need to be exact. And with fresh pasteurized milk and a clean container, YOU DO NOT NEED TO PREHEAT IT.

2. Put the starter yogurt in a cup. Again, the amount does not have to be exact, but more or less yogurt will change the needed incubation time at least a little. "Older" yogurt—say, commercial yogurt from a container that has been open for several weeks—may need a larger amount to act as quickly. (For older yogurt, I just use a heaping tablespoon instead of a level one.)

3. Add a little milk from your container to the yogurt in the cup and mix until smooth. With commercial yogurt and a spoon or fork, this can take a while, because of the stabilizers. (Gelatin is more troublesome than pectin.) But don't use a blender or frother, as that will add too much air.

Tip: Try a small whisk, rotating it in the cup by twirling the handle or rolling it between your palms. The job is done in seconds! Also, keep this tip in mind for later chapters, as it helps with dissolving all manner of stubborn powders.

4. Pour the mixed yogurt and milk from the cup back into your container and stir it into the rest of the milk. This is your yogurt *culture*—the mixture of milk and live bacteria that will become yogurt.

5. Cover your container with its lid—loosely, to let gas escape during fermenting. You might also want to stick Post-it tape on the lid, so you can note the time, date, any recipe variations, or such.

Yogurt culture in Mason jar, ready for warming

6. (Optional) If you want to wait before warming the culture, you can store your container in the refrigerator.

7. When you're ready, warm the culture with your chosen device or method. The ideal temperature for bacteria from commercial yogurt is 113°F (45°C), but you should be fine with anything between 100°F and 120°F (about 38°C and 50°C).

Be careful, though: The temperature inside the container may go higher than you've set for your warming device. So, despite this being basically a "no thermometer" method, you might want to check your yogurt temperature the *first* time and adjust the device setting as needed.

8. Keep the culture warm till it starts to firm up into yogurt. With this recipe, that will usually take about 7 to 9 hours, depending on your incubation temperature and how quickly your warming device or method got it there.

You can test for firmness by tapping the container near the top. If the yogurt is getting firm, you'll see little or no movement. You can also try slightly tilting the container. Yogurt that's starting to firm may still level off when you do that, but only sluggishly.

At this point, you could keep the yogurt warm another hour or so to get more sour. But if it's kept there too long, the liquid whey may start to separate from the solid yogurt—a sign that fermenting has passed its peak.

9. Move the container to your refrigerator, handling it gently to avoid agitation. The yogurt will continue to ferment and firm for some time as it cools.

10. Once the yogurt is chilled, just tighten the lid and leave the yogurt in your refrigerator till you're ready to eat it.

That's all! For the next batch, you can save some starter from this one. Or for greater consistency, start again with your chosen commercial yogurt. For that, you'll only need to buy a fresh tub every month or so.

Final Notes

Here are a few more things to keep in mind for this recipe.

Measurements

Really, I don't worry about using exactly three cups of milk unless I'm testing a recipe. Often, you may want more or less milk per jar to divide up a larger amount evenly, or to use up what's left in a carton. Changing the amount can change the incubation time a bit, because of the proportion of milk to starter yogurt and the time it takes to warm up the culture. But that's the only effect, and I seldom bother to even adjust for it.

The exact amount of starter yogurt is also less than critical, even to the incubation time. As I explained earlier in the book, that time depends less on starter amount than on the doubling time of the bacteria. So, you would have to increase the starter amount by about 100%, or reduce it by about 50%, before you would change the incubation time by as much as an hour.

Warming Time

My recipe calls for much less starter yogurt than most recipes do, so the time to firming can be much longer. Also, I'm starting the culture from refrigerator temperature—instead of cooling milk that's preheated—which likewise means added time for warming. (And since I don't keep my refrigerator as cold as most people, your time may be even longer!)

What if you want to shorten the time? The simplest way would be to greatly increase the amount of starter from the tablespoon I recommend. As a rough guide, double that amount for every hour you want to save. For example, use a quarter cup of starter—that's four tablespoons—to gain about two hours, or half a cup for about three.

What if you stop warming too soon and your chilled yogurt turns out not as firm or sour as you want? Not to worry! Just warm it up again, then keep it there another hour or two.

Scaling Up

If you want to make more yogurt at a time, you don't need to prepare each container individually. For example, to make yogurt from a half gallon of milk, you could mix three tablespoons of starter yogurt together with milk in a cup before pouring the mixture back into the carton or jug, shaking it, and pouring it all into jars.

In fact, if you have a container large enough for all that milk—but small enough for your warming device—you could incubate the entire amount together. Both Cambro and Carlisle, for example, make 2.7-quart clear plastic crocks—a perfect size for half a gallon. Just keep in mind that the milk might take longer to get up to temperature. Also, a wider container allows more movement of the yogurt after firming, so you might see more whey.

Carlisle 2.7-quart crock

Whey

As I explained earlier, whey is a *good* thing, as it contains much of the nutrition of the yogurt. But most of this liquid is hidden within the protein mesh you get by fermenting. If you start to see it while the yogurt is still incubating, it most likely means the bacteria have run low on milk sugar and have started consuming protein. Normally, you want to stop warming the culture *before* that point.

SMART YOGURT

When you later start spooning yogurt out of your container, it's normal and expected for some of the whey to escape the mesh and collect in the low spots. You can pour off the whey, or mix it back in, or just spoon out the yogurt and whey as is. It's all good!

Spoilage

Yogurt normally keeps longer than ordinary milk because of its higher acidity and its dominant population of friendly bacteria. And during incubation, the high temperatures used for yogurt will favor those bacteria in the first place.

But none of this protection lasts forever, and it's also possible for yogurt to be contaminated quite early. I'm not a doctor or public health expert, but here's my approach: If I see any unexpected color in my yogurt, I remove that part of it, along with a reasonable buffer around it. Then if the rest of the yogurt looks, smells, and tastes OK, I go ahead and eat it.

This, by the way, is standard advice in the world of homemade sauerkraut, where mold on top of a ferment is not uncommon. It's seldom seen as reason enough to discard a whole batch.

Some yogurt makers, though, will advise you to discard *all* the yogurt at any sign of spoilage. There's no doubt that's safer!

Optimizing Output

There are many ways to warm milk for long enough to produce yogurt. Some ways are older and have stood the test of time. Some newer ones, though, may offer real advantages.

Older Ways

Can we honor an older method while still recognizing its drawbacks? That's what I try to do in this section. But if you're using one of these methods and are quite satisfied with it, please don't think I'm pushing you to change! What's important is if it works for you.

"Warm Place"

With this method, you just place your yogurt culture somewhere that gets warmer than normal room temperature—but not *too* warm. Often this involves a pilot light, as with putting your container inside a gas oven or over a gas floor heater. Other times, the heat comes from a light bulb, like the oven light in an electric oven, or even a bulb in a lamp fixture lying in a portable cooler with the cord hanging out. Or the cooler could contain a heating pad.

This method certainly works—at least with a little trial, error, and patience. Still, you don't get the kind of control over temperature that can give you the best, fastest, and most predictable results.

Heat Insulation

This is related to the first method and can be combined with it. After getting your culture to the temperature you want, you insulate the container as best you can, using blankets, a portable cooler, whatever. You can actually buy a non-electric yogurt maker based on this principle—a device that's basically a Mason jar in a thermos bottle.

If this is done well, your culture stays warm long enough to firm up and become yogurt. But here again, you forego the best temperature control.

Electric Yogurt Maker

Many of us—including myself—started making yogurt with a dedicated yogurt maker. Most often, this is a set of small ceramic or glass cups fitted into an electrically heated base. Because the cups are small, it's fiddly to fill them all, and you can't make much yogurt at once.

Other yogurt makers have a single, larger container—but that can be less than convenient, if you want to start a new batch before you finish eating the old.

Older styles of these yogurt makers come with no controls of any kind—you just plug or unplug the device to turn it on or off. Today, some models come with digital controls for both temperature and time. This can help especially with unconventional yogurt recipes prescribing a less common temperature.

What may be more important for *any* yogurt is to maintain an even temperature throughout the container. Because an electric yogurt maker heats from the bottom, it may have trouble heating evenly after the yogurt has begun to firm up. In one popular yogurt maker's quart container, I measured a difference of 20 degrees Fahrenheit (about 10 degrees Centigrade) between top and bottom after firming.

Newer Ways

As I said, there are now ways to make yogurt that offer distinct advantages over older ones. These advantages may include:

- Precise control over temperature
- More even warming throughout the culture
- Handling of large batches as well as small
- Accepting common containers of moderate size

Sous Vide Cooker

When it comes to both precision temperature control and an even temperature for yogurt making, a sous vide portable stick cooker has to win the prize. The combination of water bath, digital control, and modestly powered heating element means you'll get extremely steady and accurate temperatures, both outside and inside your yogurt container. Another advantage is that the amount of yogurt you can make at a time is limited only by how many containers can fit in the pot or tub you use with the cooker.

To make yogurt in a sous vide water bath, place your yogurt container or containers in the pot or tub and fill it with enough water so the container is mostly submerged but doesn't float. For a Mason jar or other glass container, the water level can come right to the top of the culture. For a lighter container, it will have to stop short of that.

In some pots or tubs, getting the water level right could pose a problem: The right level for your container may be too low for running your stick cooker.

If you face that dilemma, one solution could be to add a platform to the pot or tub, raising the yogurt container. Another could be to use a taller container, such as a half-gallon milk carton.

Sous vide setup for yogurt

You could even culture the milk right in its original carton! (See the sidebar.)

Keep in mind that the water level will drop as water evaporates—but probably not by much in the few hours you'll be incubating.

The Milk Carton Gambit

For the ultimate in simplicity, why not culture your milk in its original carton? There's no need then to divide the milk among multiple containers, and when the yogurt has been eaten, the carton gets tossed!

This works best with a sous vide cooker and a water bath deep enough for you to immerse most of the carton. Also, half-gallon cartons are better than quart, making it easier to scoop out the yogurt all the way at the bottom.

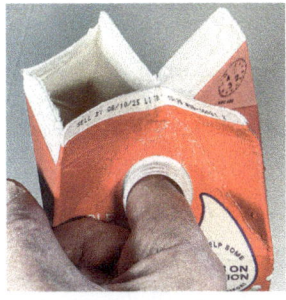

Pushing open the top

Nowadays, of course, gable top cartons come with round spouts and screw-on caps—so, the gables themselves are no longer made to open easily for pouring. Still, if you start by folding back the gables as used to be done, you can then stick a finger through the round spout and push the surfaces apart from inside. (Thanks to my wife for this tip.)

To close the carton back up, just fold the gables down again and clip the top tabs together. I've used two-inch paper clips at the ends of the tabs, while my wife favors a binder clip from above.

Later, when spooning out the yogurt, you can push the gables outward for easier access to the inside. Also, a kitchen spoon can help in reaching the bottom.

Can you heat a milk carton safely? In the old days, the cartons were waxed, so probably not. But now they're coated with food-safe plastic that withstands much higher heat than you'd use to incubate yogurt. (With a melting point above boiling, the plastic should be safe even for initially scalding the milk, if you care to—but I haven't tried that.)

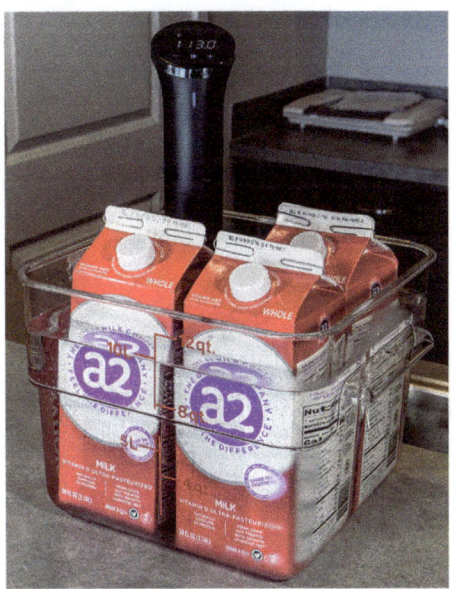

Sous vide with three milk cartons

Instant Pot

The Instant Pot has become popular for yogurt making since the introduction of a dedicated Yogurt program on newer models.

For this program, the Instant Pot's instructions tell you to pour milk directly into the inner pot. But unless you're straining the yogurt afterwards, it's much more convenient to warm your yogurt right in the containers you'll keep it in. I recommend setting up those containers in a water bath, turning the Instant Pot into an improvised sous vide cooker.

We've already discussed how to make yogurt with a sous vide stick. Using the Instant Pot for sous vide is even easier, because you don't have to worry about mismatched water levels. Just put your yogurt container or containers in the inner pot and fill it with as much water as you can without floating the container. Then set the Instant Pot to Yogurt Normal and leave it till your yogurt has set.

For the very most even heating, you can set containers on the Instant Pot's steam rack to lift them off the pot bottom. (The rack's handles can be left folded down flat.) You don't have to cover the Instant Pot—and with a one-quart Mason jar, you won't be able to, anyway.

With the Yogurt Normal setting, the Instant Pot will aim to keep its contents close to an average 40°C (104°F). That's lower than usually recommended, so fermenting is a bit slower—but since the Instant Pot is quicker than most devices at raising temperature, you save time in the initial warming.

With a six-quart model, I found I got yogurt in a single one-quart Mason jar in about seven hours, and in three jars together, in about eight. Yes, you can fit three one-quart Mason jars in a six-quart Instant Pot! You can also fit one of the 2.7-quart crocks I mentioned earlier.

Instant Pot setup for yogurt

Some even newer Instant Pot models come with a dedicated Sous Vide program. This should allow you to set the temperature to the more common 113°F (45°C) for quicker fermenting. But customer reviews I've seen for these models have complained of low reliability and long times to reach temperatures you set for sous vide and slow cooking. So, unless reviews have more lately improved, you might be wise to stick with an older, cheaper model.

Home Proofer

Now we come to my own favorite warming device: the Brød & Taylor Folding Proofer & Slow Cooker. It's a clever invention introduced only in 2011—a simple plastic box with a heating element under a metal floor, and a rack above that.

Conceptually, it's a direct descendant of the cooler-and-light-bulb contraption I described earlier, but much more elegant and efficient. It can fit up to six one-quart Mason jars or two 2.7-quart crocks—enough to make yogurt from a gallon of milk at a time! Even so, when not in use, it can fold down flat for easy storage.

In Proofer mode, it provides a temperature range of 70°F to 120°F (21°C to 49°C). Note, though, that this is from radiant heat, which means the temperature inside a glass jar will reach much higher than what you've set. Generally, I keep the proofer setting at 100°F (38°C) for yogurt. This gives me about 113°F (45°C) inside a one-quart Mason jar during incubation.

Because the proofer heats from below, you can wind up with a higher temperature at the bottom of your container after the yogurt has set, just as with an electric yogurt maker. But the difference I've measured hasn't been more than twelve degrees Fahrenheit (six degrees Centigrade)—much less than from the electric yogurt maker I tested. The higher temperature should still be tolerable for the heat-loving bacteria from commercial yogurt.

If you want to try settings lower than normal, the current generation of the proofer has a quirk you should know about. At settings of 95°F (35°C) and below, the device actually aims for a temperature five degrees Fahrenheit lower than it says! This is supposed to compensate for steam from the water tray you would use when proofing bread—but if you're *not* proofing with steam, it's just confusing. Combine this with the higher heat produced in a Mason jar, and figuring the setting you need can be truly challenging.

So, to save you some trouble, here's what I've worked out: If you want your culture around 104°F (40°C), set the proofer to 95°F (35°C) in Proofer mode. Or if you have an older model *without* a separate Proofer mode, set it to 90°F (32°C).

The Brød & Taylor may lose points on temperature control and evenness compared to the sous vide cooker or the Instant Pot—but it's still my favorite. Why is that? Because I don't have to deal with the water! Combine that with its large capacity, and the proofer easily beats anything else I've tried for convenience.

After over a decade, the Brød & Taylor is finally facing cheaper competitors—but such a proofer may have drawbacks. For instance, it may not come with the bottom rack you need for more even heating. More important, it may not reach the temperatures you want—especially if you add a rack!

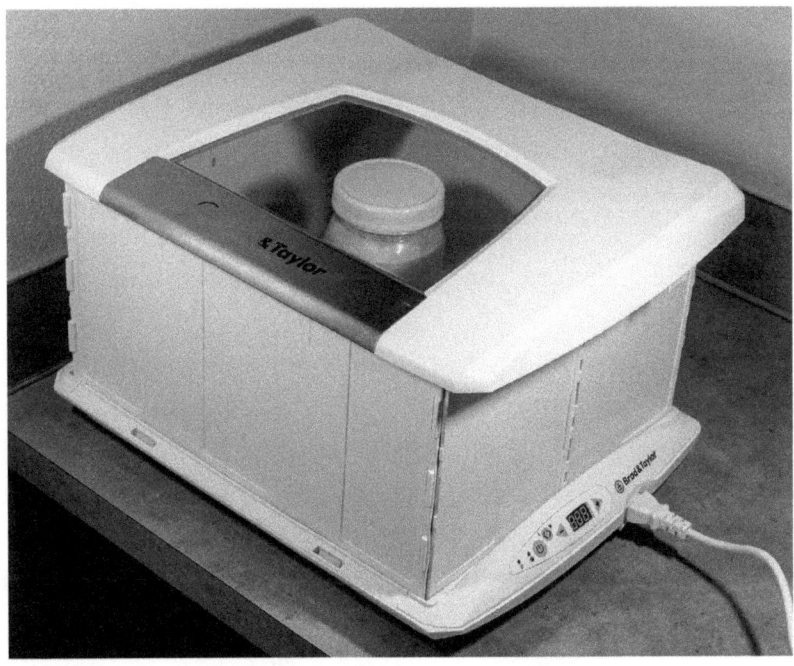

Home proofer setup for yogurt

Digital Oven

Digital ovens, or "smart ovens," have evolved quickly, especially in compact countertop models. Many now offer a variety of functions, which may even include Ferment or Proof for making yogurt and proofing bread.

If you own an oven like this, it's certainly worth trying out for yogurt making. But if you're looking to buy one, you shouldn't assume that every brand with this function will work for you. Instead, look for online reviews and posts from people who have used it *for yogurt*. You should also download the user manual for details on the function.

Here are some questions you should answer before buying any digital oven for incubating:

Does the function you want have the temperature range and precision you need? It has to hit at least somewhere between 100°F and 120°F (about 38°C and 50°C). And it's best if you can adjust the setting in single degrees, not in increments of five or ten.

Does the timer for this function have the range you need? It should allow you at least nine hours, and preferably twelve or more.

Will the temperature for this function stay fairly even? You can't have huge temperature swings like you'd get with most ovens and maybe even with other functions on the *same* oven. For a Ferment or Proof function, the heating element must turn on in short bursts only.

Will your chosen yogurt containers fit the oven? The interior height of the oven is not the same as its *usable* height. That will be several inches shorter, limited by the distance between the lowest rack or tray position and the upper heating element. With a smaller oven, this is likely to mean that one-quart Mason jars won't fit. Of course, if you can switch to shorter containers, that will solve it.

Optimum Temperature

The bacteria used to ferment commercial yogurt and most homemade yogurt are what we call *thermophilic*—"heat-loving." They're selected and bred to be most active around the high temperature of 45°C (113°F). This enables the bacteria to ferment the milk quickly, while reducing competition from microbes that prefer it cooler.

That doesn't mean you have to maintain that temperature exactly, or even within a few degrees. Every bacteria species has an optimum temperature, but deviating from it doesn't mean the bacteria will stop all activity. Within a very broad range, they will just slow it down.

You should still get yogurt as long as you keep your temperature between 100°F and 120°F (about 38°C and 50°C), or even farther afield. But for most efficient incubating of the yogurt species found in commercial yogurt, it's still best to aim for 45°C (113°F).

What if you're not starting with commercial yogurt or with yogurt made from it? Nowadays, many people are culturing single species of other lactic acid bacteria—species like L. reuteri, which is not used in commercial yogurt but *is* found in a variety of natural ferments.

For L. reuteri, you might be better off lowering your temperature a bit, to around 100°F (about 37°C)—though not every study agrees with this. In any case, if you're off from the optimum by a few degrees, that won't keep your culture from fermenting!

It's important to keep in mind that the setting on your warming device is *not* always what you'll get inside your yogurt container. For instance, many such devices work by radiant heat, and if you combine this with a glass jar, you can get much higher temperatures inside. The Brød & Taylor proofer is a good example of this. By contrast, if a device warms your container in a water bath—like a sous vide cooker does, for instance—the device setting should be almost exactly what you get inside your container.

One of the first things I do, then, with any new warming device is to test the temperatures it generates within my containers, along with how long it takes to produce those temperatures.

To do this, I pour three cups of water into the same container I plan to use for yogurt with that device, then warm it at a setting I think might work. I take the water's temperature at the beginning, and again after each half hour or hour, and I keep measuring till the temperature stays steady for at least two readings. If it's then too far from where I want, I set the device to a different temperature and try again.

All of that gives me a preliminary setting. Then, when I use that to make my first batch of yogurt in the device, I measure temperature again to make sure the setting works with the actual culture. Also, this first time, I might purposely warm the yogurt an hour longer than it needs to firm up, and then measure at different depths to make sure no part of the yogurt is overheating.

Improving Taste & Texture

As we've said, it's not easy to make nonfat yogurt with decent taste and texture. But even yogurt from whole milk can benefit from a little help in that area. So, all kinds of yogurt—nonfat, low-fat, and whole milk—often get helped along, and often in the same ways.

Some of these helpful methods are well known, while others that work as well or better may be less commonly used or discussed. Let's take a look at them one by one, noting both advantages and disadvantages. Along the way, I'll give you a few recipes based on my own recommendations.

Scalding

Scalding the milk involves heating it to around 185°F to 195°F (85°C to 90°C) for ten to thirty minutes. With slight variations, most recent yogurt recipes—other than mine—include this step as a way to make thicker yogurt. It's even built into the Instant Pot as the Yogurt High setting.

High heating like this was essential in earlier times, when it was needed to destroy microbes that would compete with the yogurt bacteria, and to deactivate enzymes that made proteins join up in other ways. But by the time yogurt became popular in Western Europe and North America, milk was almost always pasteurized, which took care of all that. So, in these parts of the world, scalding was largely unknown for home yogurt making.

Even so, scalding could be used by commercial operations, just to make yogurt thicker. To that same end, it was eventually introduced to home use by the *New York Times* food writer Harold McGee, especially in his book *On Food and Cooking*.

How does scalding work? I explained earlier how lactic acid from the yogurt bacteria makes the molecules of casein join together in a protein mesh. But there are other proteins in milk that are less affected by acid—that is, unless they're first heated above incubation temperatures.

One of these proteins is lactoglobulin. Scalding alters it in such a way that it too links up in the mesh—and the more protein in the mesh, the firmer the yogurt.

How much difference does this make? Casein is about 80% of the protein in cow's milk, while lactoglobulin is only about 10%—so, not all that much difference. Consider also that around 10% of the lactoglobulin has *already* been altered, just by the quick heat of standard pasteurization. (With UHT milk, it's around 70%.)

There's no doubt that scalding can help—but as I said before, you may only really need it with nonfat and low-fat yogurt. Otherwise, it's probably more trouble than it's worth.

Removing Whey

Straining the whey out of yogurt is how we're usually told to make Greek yogurt—or at least, what's now *called* Greek yogurt, even though traditionally most yogurt in Greece was not strained!

There's no question that straining can produce a yogurt that many people prefer. But it's a big hassle, while solutions that are less demanding may well satisfy you. And straining can be wasteful as well, as the whey contains a lot of the nutrients of the yogurt. If you're not using the whey elsewhere—say, in homemade bread or soup—those nutrients are just lost.

In any case, straining is not how most of today's *commercial* Greek yogurt is thickened. In fact, in the United States, I've found only two national brands of Greek yogurt that even claim to be strained. (That's Chobani and Fage.)

Increasing Protein

It's the protein in milk that meshes to make yogurt, and the more protein available to form the mesh, the firmer the yogurt will be. That is, the more of that protein as a *proportion* of the milk. Beyond scalding, there are several ways to boost this.

Adding dry milk powder is one common method, both at home and commercially. You can firm up your yogurt dramatically by adding up to a tablespoon per cup of milk, or three tablespoons for a three-cup recipe, using either nonfat or whole milk powder. Be aware, though, that it won't do much for taste—especially with nonfat—and you might get a grainy layer on top.

Another way to increase the proportion of protein in milk is to remove some of its water. The simplest technique, of course, is to heat the milk and let water evaporate. But though you can do this at home, I doubt you'd find it practical.

You can get a similar effect by replacing some regular milk in your recipe with evaporated milk or sweetened condensed milk. These products have been heated commercially to remove about 60% of their water. But because of stabilizers and other additives in them, I don't recommend this either. (My wife says that adding even a little evaporated nonfat milk to her yogurt made it slimy.)

Nowadays, heating is no longer the only way dairies can remove water from milk. They can do it instead through filtering. The milk they make into yogurt may already have been concentrated in this way. As another option, they might mix concentrated milk with regular.

When sold on its own in stores, milk processed by filtering is called *ultra-filtered*. By the time you buy it, water has been added to replace *some* of the water removed—but not all, leaving a higher concentration of protein than in regular milk.

The milk also typically has total sugar reduced, and it may be made lactose free besides.

Ultra-filtered milk makes impressively thick yogurt. So, it's definitely an option when yogurt thickness or high protein content is your top concern. But it does have a couple of important drawbacks.

One is taste. In my own testing, I found that yogurt made from this milk seemed almost tasteless when eaten plain. (I made it with Fairlife Whole Milk.) The difference might not bother you, especially if you're adding sweetening or other flavor ingredients. On the other hand, my wife tried the nonfat version (Fairlife Fat Free), and even *after* sweetening, she found the yogurt flat and unsatisfying.

There is a simple, fairly effective solution to the taste problem, if you're willing to do it and can tolerate it: Before making the milk into yogurt, restore lactose that was removed. I found that the taste of yogurt from my ultra-filtered milk could at least be improved by adding two teaspoons of lactose powder per cup of milk, or two tablespoons to a three-cup recipe. And though lactose is a sugar, this amount still left the yogurt nicely sour. (The lactose powder I tried was from Now Foods.)

The second drawback is not so easy to counteract. Filtering out water—just like straining out whey—removes nutrients, and not all are restored. So, if yogurt's health effects are high on your list of benefits, ultra-filtered milk would not be your best choice. (See the sidebar for more on that.)

Another option would be to replace just *some* of your regular whole milk with ultra-filtered, with or without lactose powder. This might be a good compromise between thickness, taste, and nutrition.

The concentrate used to make ultra-filtered milk is not something you can buy directly to add to your yogurt culture. You

The Lowdown on Ultra-Filtered

Yes, you can make nice, thick yogurt from ultra-filtered milk. But are you sure you want to?

Ultra-filtered milk starts with nonfat milk, which is then filtered to remove about half the water—along with most of the vitamins, minerals, lactose, and other flavor components dissolved in that half. Is this to improve the milk? No, it's mainly to cut the milk's weight and volume, so it's cheaper to ship long distance!

Once this concentrate arrives close to its final destination, pure water, fat, vitamins, and minerals are added back, in amounts needed to produce the kind of milk desired. Lactase enzyme may also be added, if the milk is meant to be lactose free.

In other words, ultra-filtered milk is not real milk, any more than reconstituted fruit juice is real juice. It's an ultra-processed, engineered, industrial product.

To sell this simulated milk, the producers typically make sure it winds up with more protein and calcium than natural milk has, and less sugar. It can then be touted as a health product! What isn't mentioned, though—and isn't even completely known—is what was removed from the milk that was never restored.

Personally, when it comes to nutrition, I trust nature over the food industry.

might imagine, though, that you could do much the same by adding a nutritional supplement made from milk's primary protein, casein.

I imagined that myself. But my test yogurt wasn't any thicker, and it tasted gritty besides. It seems that whatever way the protein was processed made it unavailable for firming the yogurt. (The casein powder I tried—which no doubt works perfectly for its intended purpose—came from Bulk Supplements.)

Adding Thickeners or Stabilizers

A common thickener like cornstarch is sometimes added to commercial yogurt. But I don't imagine many home yogurt makers would care to compromise their cultures that way.

Stabilizers like gelatin, pectin, and inulin go beyond thickeners, helping to firm up the yogurt by forming a gel that provides structure. Think of Jello or jam. You might say stabilizers provide backup to the yogurt's own protein mesh. They're especially important in commercial yogurt, because they cut down on whey separation caused by jostling in transport.

In homemade yogurt, a little whey separation is less of a concern. And since most stabilizers require heat to dissolve, they're far less likely to be worth the hassle in home use.

The only time you would likely need a stabilizer at home is if you're not making regular yogurt. Stabilizers are common, for instance, in "yogurt" from plant milks that can be soured but lack the right protein to form the mesh. (See my appendix on plant-based alternatives.)

Also, inulin has become popular from its role in the single-species yogurt recipes of Dr. William Davis. In those, the inulin can deliver an apparent success even if the desired species has died off. (Read more on that in my chapter about probiotics.)

Though stabilizers might have limited use in homemade yogurt, you shouldn't see them as harmful additives. Instead, they're natural products with health benefits of their own. Both pectin and inulin, for instance, are classed as "soluble fiber," acting as prebiotics and improving gut health, among other known benefits.

So, if you see such ingredients on commercial yogurt labels, don't think you have to avoid them. But also don't think you need them in the yogurt you make! If you do want them in your diet, there are plenty of simpler, more sensible ways to add them.

Selecting Species

Though hardly any commercial Greek yogurts are made by straining, there is something else they share in common—something the companies never publicize: They're fermented with an identical or nearly identical set of bacteria species. (Here again, Chobani and Fage are outliers, but the difference is slight.)

This set of species is selected to produce a thicker, creamier yogurt from any kind of milk. And to get the same benefit at home, all you have to do is use one of these yogurts as your starter.

Following is the set of six species. You can use this list to identify a "Greek" yogurt even if it's not called such. (For example, the same set is used by at least one brand marketing its yogurt as French-style.)

    Streptococcus thermophilus
    Lactobacillus bulgaricus
    Lactobacillus acidophilus
    Bifidobacterium lactis
    Lactobacillus casei
    Lactobacillus rhamnosus

Those are the full names, but usually you'll see the first word reduced to an initial, like so:

S. thermophilus
L. bulgaricus
L. acidophilus
B. lactis
L. casei
L. rhamnosus

The species in the yogurt's live cultures are normally listed on the container label. In the United States, though, the label may omit mention of the first two species on my list, as they're already part of the legal definition of yogurt and their presence is assumed.

Sometimes, too, you might see one or more additional species listed. The addition may or may not much affect the texture of the yogurt—but it won't likely make it thinner!

Following is a variant of my Smart Yogurt recipe, this time specifying the use of the "Greek" set of yogurt bacteria. That's the only difference, as the directions are the same, as noted.

Smart Greek Yogurt

3 cups pasteurized whole milk
1 tablespoon plain yogurt with live "Greek" cultures

With these ingredients, follow the directions for Smart Yogurt.

Adding Fat

The traditional yogurt of Greece, though not usually strained, did have one crucial difference from most yogurt today: It was not made from cow's milk at all! Sometimes it was made from goat milk, but most often from sheep milk. And a major feature of sheep milk is its extremely high fat content.

Most commercial cow's milk today is produced from Holstein cows, which give a milk with especially *low* fat content—about 4%. The fat content of sheep milk is around 8%—almost double! And in the same way that shifting from nonfat milk to whole milk greatly improves the yogurt, boosting the milk's fat content even higher gives us a food that's thick, rich, and truly delectable.

In fact, if you've never cared to eat plain yogurt—that is, yogurt with *nothing* extra mixed in—this might well change your mind!

How do you get milk with higher fat content? A simple way *might* be to use milk produced by Jersey or Guernsey cows, which may be available from smaller dairies. That milk typically has about 5% fat, as it comes from the cow. But the dairy might remove some fat for use in other products, or to appeal to weight-conscious customers. So, you'd need to check with the individual dairy for the final fat content of its milk.

With a smaller dairy, make sure its milk is pasteurized, if you don't want to pasteurize it yourself. You'll likely also want it homogenized—which is a lot harder to do at home. (In case you're wondering, sheep milk and goat milk are both *naturally* homogenized, right from the animal.)

For most of us, the most convenient way to get higher fat content is to add fat ourselves. You *could* do this with cream—called heavy cream, whipping cream, or such when you buy it in a store—but commercial cream has additives you probably don't want in your yogurt. A better choice is "half and half," a homogenized mix of equal parts milk and cream, which comes *without* the additives.

Cream you buy in stores is roughly one-third fat, so a half-and-half blend works out to a rounded 10%. Using that figure, you can easily regulate the fat content of your yogurt culture by varying how much of the recipe's milk you replace. For example, replacing one cup of milk in my Smart Yogurt recipe with one cup of half and half would yield a fat content of about 6%.

The chart below shows replacement amounts to get various fat percentages from my recipe.

What about the extra calories from the added fat? That could, of course, be a problem. But considering you'll have less need to add anything more to make your yogurt taste decent, you might find that the cost in calories is less than you'd expect. You might also consider what Mireille Guiliano says in her thought-provoking book *French Women Don't Get Fat*: If the food you eat is satisfying, you don't need as much!

Fat Content

WHOLE MILK (CUPS)	HALF AND HALF (CUPS)	FAT CONTENT
3	0	4%
2.5	.5	5%
2	1	6%
1	2	8%
0	3	10%

> # Smart French Yogurt
>
> 2 cups pasteurized whole milk
> 1 cup pasteurized half and half
> 1 tablespoon plain yogurt with live cultures
>
> With these ingredients, follow the directions for Smart Yogurt, adding the half and half to the milk at the start.

Above is another variant of my Smart Yogurt recipe, this time adding half and half. Mainly to give the recipe a different name, I call this yogurt "French," since the French are known to like their yogurt especially creamy. Here again, the only difference is in the ingredients, while the procedure is the same.

The Ultimate Yogurt?

From my recipe variations so far, you can no doubt tell that my own favored methods for improving yogurt taste and texture are to add fat and to start the yogurt with selected species. But what if you combine the two methods?

Well, you get a yogurt that's, frankly, hard to beat! This is the yogurt I've made for myself when I felt my waistline could afford it. Since it combines "Greek" and "French" methods, I call this yogurt Mediterranean.

As you can see from the photo following the recipe, it's plenty thick and firm!

Smart Mediterranean Yogurt

2 cups pasteurized whole milk
1 cup pasteurized half and half
1 tablespoon plain yogurt with live "Greek" cultures

With these ingredients, follow the directions for Smart Yogurt, adding the half and half to the milk at the start.

Smart Mediterranean Yogurt, tilted

Adding Natural Flavors

If you stop skimping on milk fat, you might find, as I do, that yogurt tastes fine all by itself. In fact, that has been my standard breakfast for years: about a half cup of plain yogurt. Still, everyone knows that yogurt with something added can be a special treat.

But it matters *when* you add that something.

For instance, you might think that adding sugar or a natural sweetener to your milk would be sure to give you sweetened yogurt. But for the bacteria in your culture, that added sweetener is food that's more easily digestible than any in the milk itself. That means it's your sweetener that will be consumed first, and it may do little more than speed up the fermenting. In the end, your yogurt may be more *sour*, not more sweet.

Other early additions can likewise have effects you don't expect or want. For example, flavor extracts, when used in typical amounts, can add enough alcohol to interfere with firming. Small bits of raw fruit too can add alcohol, produced during the ferment by the yeast they carry. (Fruit ferments, with their high concentrations of sugar, are particularly friendly to yeast—which is why they're used to make wine instead of pickles.)

Fermenting might change your intended flavor to one truly awful. Or something in the flavor ingredient might curdle the milk long before the yogurt firms up. Even if an ingredient is fine in the ferment, it might not stay dispersed evenly enough in the milk during incubation. A powder might sink to the bottom; an oil might rise to the top.

Generally, then, the simplest, safest way for you to add flavor ingredients is to wait till your yogurt is firm and chilled—or maybe even till you serve it. Once yogurt is cold, the bacteria won't react much with added ingredients, and the firmness of the yogurt will hold them in place.

With that said, there *are* ways to avoid some of these problems and add flavor to your milk *before* incubation. And there are good reasons you might want to.

Sometimes it might simply be easier to add the ingredient to milk than to firmed yogurt, whether that's just after cooling or right before serving. But flavored yogurt can also invite uses you might not otherwise consider.

For instance, sweet-flavored yogurt might serve as a topping for dessert, substituting for whipped cream—especially if you replace some or all of the milk with half and half. Or a savory yogurt might replace sour cream as a topping for baked potatoes.

Yogurt of either type might serve as a dressing—or the base of one—for salads, grains, or vegetables. Even whole milk yogurt should give you fewer calories than a typical low-calorie salad dressing.

Adding flavor ingredients to your milk can also double as a way to give yogurt attractive coloring. A standard white yogurt might lack visual appeal for either adults or children—but what about a light orange or purple?

We'll limit ourselves here to *natural* flavors. Besides helping narrow the topic, this removes a whole range of concerns about possible dangers. Not that natural ingredients can't come with their own issues!

Sweetening with Lactose

Some yogurts with added flavors can be eaten straight, while others call for sweetening to keep them from being too sour or too bitter. If you're eating such yogurt on its own, you can always add sweetening before serving. But for toppings and dressings especially, it can be most convenient to sweeten the milk before fermenting.

But how to sweeten it then, if any kind of sugar just feeds your yogurt bacteria? One way is to add lactose powder. This native milk sugar is less sweet than other sugar types—about 20% to 40% as sweet as sucrose (table sugar)—but it also breaks down less readily as the milk ferments, so more of it is likely to remain afterwards.

Of course, adding lactose is a very *bad* idea if the yogurt is for anyone who is lactose intolerant! Some people can eat yogurt only because most of the lactose has been fermented out. And even people who can normally tolerate lactose may have trouble digesting too much.

So, how much to add? The amount of lactose in milk before fermenting is about 12 grams (about half an ounce) per cup—about equivalent to one tablespoon of lactose powder. In normal yogurt making, around two-thirds of that lactose might be consumed by the yogurt bacteria.

Say, then, you want your yogurt to come out about as sweet as the original milk, as a way to balance the sourness. I found I could get a fairly neutral taste by adding two teaspoons of lactose powder per cup of milk, or two tablespoons for a three-cup recipe. (For roughly the same effect with a lactose-free milk, I needed one tablespoon of lactose powder per cup of milk, or three tablespoons for a three-cup recipe.)

You can adjust from there. For an added touch of sweetness with regular milk, make it a full tablespoon per cup, or three tablespoons for a three-cup recipe. For a touch of sourness, use a smaller amount. (The lactose powder I tried was from Now Foods.)

Be aware, though, that the reduced sweetness of lactose does *not* come with reduced calories! So, if you want your yogurt *really* sweet, adding enough lactose powder may not be the best way to get there, if you care about your weight.

My lactose powder dissolved easily, so I was able to add it directly to the cold milk in my yogurt jar. But if yours gives you trouble, see the sidebar "Dissolving a Powder."

Flavoring with Vanilla

Vanilla is a natural as a flavoring for yogurt. And when most people think of adding vanilla, they think of vanilla extract. But as I mentioned, the amount of alcohol such an extract would typically add to your milk will interfere with firming. That's a shame, because adding vanilla early is by far the most convenient way to include that flavor.

Luckily, there are several alternatives to vanilla extract.

Vanilla Flavoring

Though *extracts* are all alcohol-based, you can buy vanilla *flavoring* (or *flavor*) with a base of glycerin instead. The one I tested worked very well for yogurt. (It was Watkins Organic Pure Vanilla Alcohol Free Flavoring.)

Look for vanilla labeled "alcohol free" or "non-alcoholic"—but also check the ingredients on the label. Ideally, they should be limited to glycerin, water, and vanilla bean. (Note that *glycerin*, *glycerine*, and *glycerol* are the same or close enough, but *glycol* is entirely different.)

If other ingredients are listed, they might or might not work in your yogurt. You might have to test to know for sure. Some questionable ones might be present in such small amounts that they won't matter.

Glycerin can be made from plants, animals, or petroleum. Most of the glycerin used in food is from plants—but if you

Dissolving a Powder

A number of ingredients recommended in this chapter come as powders, "instant" or otherwise. Some of these will dissolve in almost any liquid, but many need warmth, refusing to dissolve properly in cold milk. Luckily, your milk generally needs to come only to around incubation temperature.

Here are several ways you can work this into your usual procedure for yogurt making:

• Wait till your milk-and-starter culture has reached incubating temperature before adding the powder.

• Warm your milk quickly to incubating temperature before adding both powder *and* starter.

• Before adding starter, pour a half cup of your milk into a cup and quickly warm it to incubating temperature. Mix in the powder, then pour this mixture back in with the rest.

This final method is the one I prefer, as the powder may be easier to first mix into a smaller amount of milk. From testing, I know my microwave takes about thirty seconds to warm half a cup of milk from refrigerator temperature to incubating temperature—so, I don't need a thermometer to check that. Also, since I haven't yet added the starter, I don't need to worry about overheating.

If you have more than one powder to dissolve, a half cup of milk might not be enough for it all. But you can always mix in just some of the powder, pour the mixture back in, then pour out another half cup to mix more.

As I warned before, you don't want to use a blender or frother on the milk, because it will mix in too much air. But a whisk works wonderfully on powders, just as it does on yogurt starter. Again, you can rotate a small whisk inside a cup, either by twirling it with the fingers of one hand or by rolling the handle between your palms.

One-handed twirl

Two-handed roll

want to be sure about yours, look on the label for something like "vegetable glycerin" or "organic glycerin," or contact the manufacturer.

Generally, you would use the same amount of flavoring as you would of extract. That would typically be a half teaspoon per cup of milk, or 1½ teaspoons total for a three-cup recipe.

Vanilla Powder

Also called ground vanilla, *t*his is another, less familiar alternative. I found it to be a flavorful, convenient, and economic way to get my vanilla. (The brand I tried was Talcufon Vanilla Bean Powder.)

Modern brands of vanilla powder are 100% vanilla bean. Note, though, that older brands can include sugar—in fact, they can be *mostly* sugar. So, check the label.

If you search for advice on how much powder to use compared to extract, recommendations vary from an equal amount to only half—possibly based on whether that's meant for pure vanilla or vanilla with sugar. For pure vanilla powder, I recommend an amount somewhere in between: one teaspoon of powder for a three-cup recipe.

When adding a powder to milk, you can't always see how well it's dissolving—so, to check my vanilla powder, I tried a teaspoon in three cups of cold water. While other brands might act differently, this powder dissolved promptly into a lovely translucent brown, leaving no residue or even cloudiness in the water. For yogurt, you can't get better than that!

If yours doesn't dissolve as well, see the sidebar "Dissolving a Powder."

Vanilla Beans

Though not the most convenient alternative, you can soak whole vanilla beans in milk, warm or cold, to draw out their flavor. I found that a single bean was enough to give good flavor to a three-cup recipe, and the bean could be reused too. Not to mention, the open package made my kitchen smell great! (I tested with Madagascar Vanilla Beans from Vanilla Bean Kings.)

Flavoring with Juices

Fruit and vegetable juices provide a quick, easy way to add flavor to your milk before fermenting. Possibilities include apple, pear, grape, mango, carrot, tomato . . . Well, the possibilities are almost endless, in both single flavors and blends. And if you can't buy a juice you want, you can make your own!

Just as most juices should be shaken before drinking, you might want to stir up your juice-flavored yogurt once it firms. Otherwise, the flavor might be weaker toward the top, stronger toward the bottom. This difference might not bother you, though.

Also, yogurts with fruit flavors are often good candidates for pre-sweetening with lactose powder as described earlier. This can take the acid edge off a normally sweet flavor, or keep another flavor from becoming *too* tart. But many flavors do fine without adjustment, and your own taste may vary.

Besides coming in so many flavors, juices come in many forms—so, you'll want to make sure you understand what you're using. That means reading labels of any juice you buy. Otherwise, for instance, you might wind up with a "juice product" with very little juice! (See the sidebar "Juice—A Buyer's Guide" for a brief review of juice terms.)

Juice—A Buyer's Guide

Juices come in so many forms, it's helpful to define some general terms. Just keep in mind there are always variations in the ways juice is processed.

Fresh juice is the kind of unprocessed juice you can make at home, whether by squeezing, juicing, or blending, with or without the pulp strained out. If made and sold commercially, it must be refrigerated.

Pure juice is juice sold commercially with nothing removed or added. But it must still be pasteurized to let it be stored, packaged, and sold without refrigeration.

Juice concentrate has first been pasteurized and then had much of the water filtered out—along with some flavor and nutrition. This allows more efficient transport and storage. The concentrate you can buy directly is normally frozen, though it can also come in bottles, especially for less common flavors. Concentrate may have citric or ascorbic acid added to help make up for what was removed.

Reconstituted juice is concentrate with water added back. This is what most often passes for juice in stores today. You won't find "reconstituted" in the product name, but on ingredients labels, this juice is identified by the words "from concentrate."

Juice powder is pure, pasteurized juice that has been freeze-dried or spray-dried. Nothing but water has been removed, though the heat of drying will have its own effects.

Juice drink. These are sugary drinks that are less than one-third real juice. Not recommended for human consumption!

Sometimes it's obvious what you're getting, sometimes not. Frozen juice is almost always a concentrate. Refrigerated juices are generally pure. But other bottled juices can be pure, reconstituted, or a combination of those. You really need to read labels to know what you're buying.

Another reason to check labels: Just because a product is "100% juice" doesn't mean it's 100% the juice in the product name! Ingredients on the label are listed in descending order of amount included, and you may well find a cheaper, more common juice in second place, or even in first.

You also need to read labels to check for thickeners like starches and for stabilizers like gums. At least some of these can disrupt the firming of your yogurt, so you should avoid them.

You may not be bothered by a bit of pulp or other residue sinking to the bottom of your culture, but you likely want to avoid too much. Some juices like apple can be bought with little or no pulp, and concentrates often come without. But with other juices, you may need to strain them yourself.

Of course, you'll want to avoid juices that can curdle your milk. If you're not sure about a juice, you can often find an answer in an online search. But even a juice known to curdle milk might be safe enough if you mix it with your milk at refrigerator temperature, or if you just add less of it.

One common cause of curdling in milk is citric acid, found naturally in citrus fruit juices and sometimes added to others. The more acid, the worse the curdling. So, you can pretty much forget about lemon or lime juice—but you might get away with a little orange.

With some fruits, like pineapple and kiwi, milk curdling can instead be caused by enzymes—and some or all of those might be destroyed by pasteurization. That means some bottled juices and concentrates will work better than fresh! I've had no trouble, for instance, flavoring yogurt with frozen pineapple concentrate.

Juice can be added to your milk before or after you add starter. The amount, though, is limited by how much the yogurt's protein mesh will be able to hold without breaking down or leaking. This will vary according to the strength of the mesh and the thickness of the yogurt, as affected by factors discussed in the previous chapter. But keep in mind that adding juice will itself weaken and thin the yogurt by diluting the milk's protein and fat.

Let's say you're making yogurt with whole milk, with a commercial Greek yogurt as starter, in a three-cup recipe. With fresh, pure, or reconstituted juice, you might then add up to a half cup to your milk. Trying to add more would likely not be a good idea, since *even* with half a cup, you might start to see cracks in the yogurt surface.

One way to bypass this limit would be to use ultra-filtered milk, with its extra protein strengthening the mesh. For example, with Fairlife's 50% more protein, I found I could substitute a full cup of juice for a full cup of milk in a three-cup recipe. You might also find that the added juice is enough to mask the typically flat taste of such milk—but if not, you could fix it next time by also adding a tablespoon or two of lactose powder.

With juice concentrate, the limit would be more like a quarter cup—even with a milk like Fairlife. Beyond that amount, the juice can leak badly from the mesh and start to pool in dark spots—though that may or may not bother you.

Just as with regular juice, thicker parts of concentrate can settle. That's something you don't need to think about if, say, you're emptying a can into a pitcher. But if you're pouring out just a portion for yogurt, it means you must first stir or shake the contents!

Juice powders are an interesting alternative, because they add *no* liquid to dilute the milk, and also because they come in a variety of common and uncommon flavors. With these powders, you might add two teaspoons to a cup of milk, or two tablespoons for a three-cup recipe. But they will almost certainly need to be mixed into milk that's warm—so, see the sidebar "Dissolving a Powder" for tips.

Once you start using juices in yogurt, it's hard to resist blending flavors. This is a good way to create theme yogurts. A Tropical yogurt, for instance, could be made with mango, papaya, and guava juices or juice powders. A Garden yogurt could include carrot and tomato juices, plus juice made from garden greens.

In fact, that Garden yogurt—made from juices left over from my experiments—is exactly what I made for my wife, to use as a base for salad dressing. For the milk, I chose Fairlife, so I could include a full cup of juice. To this yogurt of mine, she added lime juice powder (left over from an experiment that had curdled my milk), lemon juice powder, sour salt (citric acid), and Green Goddess herbs.

At least, that was *one* day's recipe. Her additions changed from day to day!

Flavoring by Infusion

Infusion is a fancy name for a tea, as well as for how it's made. Usually, infusions are made by soaking a flavor ingredient in water that has been brought to a boil. But you can replace water with milk, and you don't need to heat it nearly that high.

Likely candidates for milk infusions include herbs, spices, and seeds. You may also succeed with fresh or dried fruits and vegetables. Also possible are roasted "beans" like coffee and cocoa—either whole or coarsely ground—as well as whole vanilla beans, as I suggested earlier.

Be aware, though, that not all flavors will survive fermenting. My tests with roasted cashews and roasted peanuts, for example, wound up tasting awful. You'll just have to experiment.

There's no one right way to make an infusion for yogurt. Basically, though, you want to warm the milk, soak the flavor ingredient long enough for the milk to pick up the flavor, then strain out the ingredient. Here's how I usually work that into my own yogurt making:

1. Add the flavor ingredient to the milk, right after adding the starter.

2. Place the mixture in the warming device and warm as usual. (I might stir or shake the mixture now and then to help disperse the flavor.)

3. Once the mixture has reached normal incubating temperature, strain out the flavor ingredient and continue incubating as usual. (In my proofer, reaching that temperature takes about three hours, but your time might be less.)

Depending on the flavor ingredient, a number of variations on this method should work. Some examples:

- Let the ingredient soak longer before straining—but just make sure you remove it before the yogurt firms! (You might need to do this if your warming device brings the milk to incubation temperature quickly.)
- Wait to add the yogurt starter till you've strained out the flavor ingredient. (This would reduce the chance of unexpected results from fermenting the plant.)
- Infuse the flavor in the milk without adding starter, then refrigerate the milk and wait to make yogurt later. (In this case, you might find it simpler to warm the milk in a microwave for the infusion.)
- Infuse at a higher temperature, even bringing the milk briefly to a simmer. (But then you'd have to deal with cooling before adding starter, skin forming on the milk, and so on—and none of this should be needed.)

Things can get trickier if you're also dissolving a flavoring powder into the milk. You might want to wait to add that till the milk has warmed and you've already strained out the infused ingredient. But like I said, there's no one right way!

How much of an ingredient you should add to your milk will depend largely on what kind it is, along with its brand, age, soaking time, preferences of your own, and so on. But to give you starting points to figure from, here are some amounts I worked out in my own testing:

- For whole spices like caraway or cumin seeds, dried leaves like peppermint, or dried roots like ginger, I used one teaspoon per cup of milk, or one tablespoon for a three-cup recipe.
- For roasted/toasted ingredients like sesame seeds or cacao nibs, two teaspoons per cup, or two tablespoons for a three-cup recipe.
- For bits of dried fruit or grated vegetables, half a cup for a three-cup recipe.

Many of the ingredients you might want to infuse are available, at least in one form, as a finely ground powder—one that does *not* dissolve. You might be tempted to use them like that, and this will certainly give you the flavor. The problem is, how do you remove the powder once you're finished infusing? Though an ultra-fine strainer might be able to filter out most of it, that strainer is also likely to clog quickly.

So, I recommend sticking with forms of the ingredient that are easier to strain out. For example, if you're infusing ginger, use dried root instead of ground ginger. Or for coffee beans, choose the coarsest grind you can, or even grind the beans roughly in a blender. Even with such ingredients, you might want to sift them before infusing to remove most of the finer particles mixed in.

With fruit, avoid leaving it in the milk for much longer than I recommend. Because of fruit's concentrated sugar, bits of it can shift the ferment toward the alcoholic if left in too long.

As when adding juices, you have to take care about curdling. Fresh ginger root, for example, will curdle the milk very quickly, even though the dried root works fine. Again, search online if you're not sure.

Speaking of curdling, infusing can solve the problem of milk with citrus. Instead of adding the fruit's juice, grate the zest—the colored part of the rind—and infuse the citrus flavor from that. Just make sure your fruit is organic. Since citrus rind isn't normally eaten, farmers commonly spray that fruit with stronger pesticides or more of them.

You'll need about one teaspoon of grated zest per cup of milk, or one tablespoon for a three-cup recipe. (Just to give you an idea, you'd need about one lemon for a teaspoon of grated zest, or three or four for a tablespoon.) Avoid grating the white part of the rind, since it's bitter.

As with juices, it's easy to blend different spices and herbs you're infusing. This is a good way to create yogurts with national and regional themes. For an Italian yogurt, for instance, you could infuse basil, oregano, and rosemary. A Middle Eastern yogurt might include flavors of cumin seed, coriander seed, and toasted sesame seeds.

Flavoring with Beverage Powders

Powders are available for making many beverages, and these powders often work in yogurt as well. Among them are the juice powders I've already mentioned, produced from regular juice by freeze drying or spray drying.

The same drying methods give us instant coffee and instant tea—regular or herbal—made from infusions or brews. These powders too may work fine in yogurt, while saving you the trouble of infusing ingredients yourself. Here again, though, you'll often need to mix them into warm milk, as discussed in the sidebar "Dissolving a Powder."

For coffee yogurt, I've had good results with several brands of instant espresso and regular instant coffee, using one teaspoon of powder per cup of milk, or one tablespoon for a three-cup recipe. (Higher amounts interfered with firming, and even this amount was pushing it with the regular coffee.)

Though most instant beverage powders are produced by drying, one exception is hot cocoa mix. Besides the ingredients all being mixed together for convenience, the powders are also "agglomerated" to help them dissolve more easily, especially in *cold* liquids. Most brands, unfortunately, come with additives that might interfere with yogurt firming. (So does store-bought chocolate milk, which is why I'm not saying any more about it.)

When I tested mixes *without* such additives, I wound up with sludge at the bottom of my yogurt jar, even when mixing into warm milk. But that sludge was pure chocolate, so you might not mind it! Also, the yogurt *above* that sludge tasted quite nice, with vastly better flavor than my test infusions from raw or toasted cacao nibs.

The amount I used for cocoa mixes was two teaspoons per cup of milk, or two tablespoons for a three-cup recipe. (I tested Ghirardelli Double Chocolate Hot Cocoa Mix and Starbucks Double Chocolate Hot Cocoa.) And the absolute best results I got taste-wise were when I substituted half and half for *all* the milk. Decadence!

If you'd rather avoid commercial mixes, you can start from scratch with the same amount of cocoa powder plus vanilla and lactose powders. But you will definitely need warm milk to dissolve the cocoa powder, unlike with instant hot cocoa. (Whatever you do, though, you still get the sludge.)

Starting from scratch may be less convenient, but it has at least one advantage: You can replace the instants' Dutch-process cocoa with natural cocoa, which is richer in both flavor and nutrients. Though Dutch-process cocoa is said to dissolve more easily than natural, I haven't noticed that difference in pure cocoa. (I tested with Hershey's 100% Cacao Natural Unsweetened Cocoa and, for the Dutch process, Hershey's 100% Cacao Special Dark Cocoa.)

Not sure if your cocoa powder is natural or Dutch process? Read the ingredients on the label. If it just says "cocoa," it's natural. Dutch-process cocoa will always be called something like that, or like "cocoa processed with alkali."

Beverage powders too are fun to mix, both among themselves and with different kinds of flavoring. Add some vanilla powder to a coffee yogurt and get vanilla latte yogurt. Add to that some hot cocoa mix or your own cocoa mixture and get caffè mocha yogurt. Add to *that* some peppermint flavoring—with either an infusion of peppermint leaves or a drop or two of peppermint extract—and you have peppermint mocha yogurt. Then maybe some strawberry juice powder on top of that?

Who needs Starbucks!

Adding Natural Colors

Natural flavors are a fun way to introduce color into yogurt. In fact, they can give you the complete rainbow of primary and secondary colors—or at least softened versions of them. Be warned, though: The strongest color doesn't always come with the best flavor!

Here's how I achieved the color shown in each photo, including whether I used regular whole milk or the ultra-filtered Fairlife—for extra protein to strengthen the yogurt mesh and discourage "leaking" of colored whey. All samples were started with commercial Greek yogurt.

Grape

Beet

Purple. This is probably the easiest color, as the juices of most so-called red and blue fruits are actually shades of purple. Though some of these juices will curdle milk, I can vouch for grape, whether from bottled juice or concentrate. Not only will it provide good color without curdling, it also tastes great, with or without sweetening. In fact, grape yogurt is one of my favorite discoveries from writing this book! (For my photo: one-quarter cup grape juice concentrate in three cups regular milk.)

Red. If you're thinking tomato, forget it. It never gave me better than pale pink, even with tomato concentrate—aka tomato paste. What finally *did* work was beet! I had great results from both beet root powder and bottled beet juice, despite its added lemon. I wouldn't eat this yogurt straight, but it could work as salad dressing. (For my photo: one-half cup bottled beet juice in three cups Fairlife.)

Orange. Another easy color—as long you like carrot juice! *Fruit* juices of this color are generally too pale, and orange juice itself can cause curdling. (For my photo: one cup bottled carrot juice in two cups Fairlife.)

Carrot

SMART YOGURT

Turmeric

Yellow. Turmeric was the answer here. I infused "cut and sifted" dried turmeric root, commonly used for turmeric tea. Personally, I found the yogurt distasteful—possibly from using too much of the turmeric. Later, though, I got a more palatable but still colorful result from Golden Milk powder, a popular instant drink that blends turmeric with spices like ginger and cinnamon. With a little sweetening, that yogurt was quite tasty. (For my photo: one-half cup dried turmeric root infused in three cups regular milk.)

Wheatgrass

Green. The only strong green I could get was from wheatgrass juice powder. Frankly, I found the stuff obnoxious—to taste, to smell, and to work with—but for the sake of its health benefits, some people do tolerate it. By the way, I got equally strong color from the equally obnoxious spirulina powder—but it was more of a blue-green. (For my photo: two tablespoons wheatgrass juice powder in three cups Fairlife.)

Blue tea

Blue. The one solution here was butterfly pea flowers, the source of what's called blue tea. An infusion gives your milk a brilliant blue, though the yogurt's acidity later nudges it along the spectrum. Go easy on the flowers, because too much of them can give a horrible taste and smell. The yogurt shown here was tolerable but still not tasty. (For my photo: two tablespoons dried butterfly pea flowers infused in three cups regular milk.)

Coffee

Besides the rainbow colors, you have shades of brown—but they're no challenge at all. Just add instant coffee, or cocoa, or both! (For my photo: one tablespoon instant espresso in three cups regular milk.)

Reducing Intolerance

Have trouble digesting milk? Then yogurt, with or without special treatment, could be the way to finally get dairy into your regular diet. But you'll want to consider the different ways to make this work for you—and also make sure you know exactly what it is you can't tolerate!

Lactose Intolerance

Most of the sugar naturally found in milk is in the form of *lactose*, a complex form of sugar that must be broken down into simpler forms for our bodies to use it. The digestive enzyme that breaks it down is called *lactase*. (Yes, the spelling is the same, apart from an *a* replacing the *o*.)

Just to be clear, lactose is far from a villain. In some ways, it can be more healthful than simpler sugars. This is especially true for infants, who generally thrive on it.

But while adults usually produce at least *some* needed lactase, some of us may not produce enough to digest all the lactose we consume. The undigested lactose then reaches the intestines, where it feeds the bacteria found there. If the bacteria become overactive, you can get gas, diarrhea, and other discomforts. A shortfall in lactase production can be temporary, or it can be permanent, especially as we get older.

If you're suffering from such *lactose intolerance*, you can likely still eat *some* dairy. But you'd probably want to limit your portions, or take lactase supplements with your meals, or choose products with reduced lactose.

Products with *greatly* reduced lactose are called *lactose free*. In reality, you can never remove *all* lactose from dairy—you can only reduce the amount to where it's highly unlikely to cause trouble. For this reason, the U.S. Food and Drug Administration permits the "lactose free" label on any food with no more than half a gram of lactose per serving.

Just in itself, yogurt is a pretty good choice of dairy if you're lactose intolerant. That's because, as the yogurt bacteria ferment the milk, they themselves break down much of the lactose to get simpler sugars to feed on. This reduction may be enough to let you handle what's left.

Still, when making yogurt, there are several ways to reduce the lactose much further, to earn that "lactose free" label.

Using Lactose-Free Milk

The simplest way to make lactose-free yogurt is to start with lactose-free milk. That's what I've done with my first recipe here.

Lactose free milk itself is made in a couple of different ways. One is to just add the lactase enzyme to it, breaking down the lactose into simpler sugars, right there in the milk.

Lactose can also be also removed when milk is *ultra-filtered*. This filtering initially removes about half the water, along with about an equal portion of the lactose. When the milk is later reconstituted, that lactose is not replaced, and the lactase enzyme is typically added to break down the lactose that remains.

You may wonder how lactose reduction would affect your yogurt making, since the lactic acid bacteria normally rely on the milk's lactose for food. The answer is, not as much as you might expect. The bacteria don't need the lactose itself, they need the simpler sugars they get from breaking it down. The lactase enzyme, then, is doing some of their work for them, which may actually speed up fermenting.

In the ultra-filtered milk, it's true there's also less *total* sugar for the bacteria to feed on. But in regular yogurt making, the bacteria use only a portion of the milk's sugar—so, after filtering, there's still enough sugar left to feed them.

In my own testing, I had no trouble making yogurt from either kind of lactose-free milk. And as I discussed in my chapter on taste and texture, the ultra-filtered milk yields yogurt that's thicker and richer than most. Still, for reasons of taste and nutrition—also discussed in that chapter—I would recommend milk processed only with lactase. (I used Lactaid as a sample of milk processed only with lactase, and Fairlife as a sample of milk that was ultra-filtered.)

Two brands of lactose-free milk

For this recipe—if you want to get really strict about ingredients—you could also find lactose-free yogurt for your starter. But the amount of lactose in a tablespoon of yogurt shouldn't be enough to worry about, and the fermentation will reduce that amount anyway.

Smart Lactose-Free Yogurt #1

3 cups lactose-free pasteurized whole milk
1 tablespoon plain yogurt with live cultures

With these ingredients, follow the directions for Smart Yogurt.

Adding the Enzyme

The second way to make lactose-free yogurt is to start with regular milk but then add the lactase enzyme yourself. This is what I've done in my second recipe. (I tested with Nutricost Lactase Powder.)

The enzyme is active all during incubation, then keeps working after the milk is chilled, for as long as it's stored. Lactase never gives up or gets used up—so, eventually, your yogurt *will* be lactose free. The drawback to this method, though, is you can't tell for sure *when* your yogurt reaches that point.

According to one commercial maker of lactose-free yogurt, adding the lactase enzyme makes its yogurt naturally sweeter, so that not as much sweetener needs to be added. Maybe this is true in the long run, but in my testing, I found the lactase at first made the yogurt more *sour*. I figured this was because the enzyme provided the bacteria with more easily accessible food, boosting their production of lactic acid.

Smart Lactose-Free Yogurt #2

3 cups lactose-free pasteurized whole milk
1 teaspoon lactase powder
1 tablespoon plain yogurt with live cultures

With these ingredients, follow the directions for Smart Yogurt, adding the lactase powder to the milk at the start.

Extending the Ferment

Still another way to get lactose-free yogurt is to just keep incubating it and let it ferment—some say for 24 hours. In fact, "24-hour yogurt" is part of two popular diets for gut health: the Specific Carbohydrate Diet (SCD) and the Gut and Psychology Syndrome (GAPS) Diet.

After trying this, I can readily believe the yogurt is lactose free, because it's also extremely sour! You probably won't like it unsweetened, to put it mildly. Truly, with this level of acidity, you might start worrying about your tooth enamel.

Also, contrary to claims, the probiotic value of such yogurt would be *diminished*, not enhanced. That's because most of the bacteria would be killed off by starvation, or by all the lactic acid they're now steeped in, or by both together.

Do you get that I don't recommend this method?

Tip: Whenever you see claims of high bacteria counts in yogurt fermented for long periods, try to determine whether these counts are based on theoretical calculations or on lab tests—and if on tests, whether they counted only live bacteria, or live and dead together. (More on this in the next chapter, when we look at some popular recipes for cultivating single bacteria species.)

Ditching Dairy

As a final method to cut lactose, you can opt out of dairy entirely and switch to vegan yogurt. I'll explore this option further in my appendix on plant-based alternatives.

Protein Intolerance

Strangely enough, not everyone who is "lactose intolerant" is lactose intolerant.

Say what?

If you seem to be lactose intolerant, you might instead be reacting to casein, the primary milk protein. The symptoms can be very like those of lactose intolerance, making it less likely that a protein reaction is recognized or even considered.

If you simply can't handle casein, then obviously, dairy yogurt is not a good fit for you. On the other hand, you might not react the same to *all forms* of the protein.

There are several types of casein, the main ones being called A1 and A2. While A1 casein is the more common type in milk from modern dairying, some researchers believe we humans are better equipped to digest A2. Whether or not this is true in general—and just be aware, it's widely disputed—it might still be true for some individuals or ethnic groups.

So, if you have symptoms of lactose intolerance that persist with lactose-free milk, you might try milk produced with A2 casein only. And in fact, you might try it anyway, as A2 milk has been linked to other advantages as well—from health benefits to better taste to improved animal welfare.

There is actually one company that markets such milk internationally: the A2 Milk Company. Its milk can be found in many supermarkets. (Not in the UK, though, as of this writing.)

If you can't find milk clearly labeled A2, you might look for milk from a smaller, local dairy with Guernsey or Jersey cows. Such dairies will often feature A2 milk, and they may say so on their website—or you can contact them to ask.

Friend or Foe?

Lactose is often discussed as one of the FODMAPs, an acronym for a group of carbohydrates that may be difficult or even impossible for us to digest. Such nutrients can pass unaltered through our systems till they reach the intestines, where there are bacteria that can and do digest them. This bacterial activity can result in excess gas, diarrhea, and other discomforts.

Wait a minute. Does this description of FODMAPs sound familiar? It should, because it's also the description of a *prebiotic*. That's a nutrient that passes through the digestive system to the intestines, where it feeds the bacteria of the microbiome, the source of many health benefits.

Hold on. Doesn't that also sound a lot like *dietary fiber*, which we've been taught is important for regular bowel movements? Well, yes, because prebiotics are basically one kind or another of dietary fiber—and so are the FODMAP food types, along with lactose and some other sweeteners. In fact, lactose itself is sometimes called a "potential fiber," in being harder to digest for some people.

Then which is it? Are these substances good or bad for you? Well, that depends on which health system you're being sold. Fearful of IBS or SIBO or "wheat belly"? Avoid FODMAPs! Eager to nurture the microbiome? Consume prebiotics!

What's really fun is to watch advocates of one system trying to tell you how to avoid FODMAPs and *at the same time* get enough prebiotics and fiber—or how to get your prebiotics and fiber while avoiding FODMAPs. Breathe, don't breathe!

I'm not a doctor, but I believe that eating a varied, balanced diet of natural, whole foods, while avoiding large amounts of any foods giving us obvious discomfort, will provide the amount of fiber/prebiotics/FODMAPs that meets our needs. If, on the other hand, we try to patch the modern, industrialized diet with one faddish diet change after another, it's never going to work—because no one will ever know enough to tell us everything we should or shouldn't be eating.

Boosting Probiotics

Yogurt with live cultures is often touted as a great *probiotic*—a food that delivers beneficial microbes to your digestive system. So, three cheers for yogurt, right?

Well, no. More like one or two. If you buy Greek yogurt, it will have only about six bacteria species. If it's *not* Greek, it may have only two. Meanwhile, a *naturally* fermented food will have *dozens*. So, among probiotics, store-bought yogurt does not rank that high.

But it's possible to make your own yogurt that *does* contain a wide range of probiotic bacteria—and in ways that are easier and more effective than any you're likely to have heard about. This chapter will lay out several approaches I've developed on my own, drawing on my experience with other natural ferments.

Is It Worth It?

But first—at the risk of undercutting my own contribution—I have to ask: Is this really what you want to do?

Many people now know of the importance of the *microbiome*—the gut's constellation of microbes that symbiotically help our bodies to function and maintain health. But those same people may have an entirely wrong idea about how to nurture it. They may think of these microbes like vitamins, believing that the greater quantity they ingest, the healthier they'll be.

The microbiome doesn't work like that. Once a microbe is introduced into the gut in a reasonable quantity and establishes itself, it's *there*. A healthy body with a good diet will maintain it perpetually, ideally in proper balance with all the other beneficial microbes in the system. The microbe's numbers will fluctuate with conditions in your gut, but the microbe won't need to be reintroduced—unless you kill it off with antibiotics or an impoverished diet.

In fact, eating greater amounts of food with that microbe isn't even a good way to increase the population, since most of the microbes you take in are killed off by stomach acid. To build that population, you'd do better to just eat more fiber—*prebiotics*, in today's lingo—since that's what gut microbes love to feed on.

As we've discussed, yogurt is fermented by lactic acid bacteria, as are natural plant-based ferments like sauerkraut, pickles, and sourdough starter. And we're not talking about the same *family* of bacteria, we're talking about the same exact species!

Out of curiosity, I made a list of the bacteria species most commonly used to make commercial yogurt. I then found that *every one of them* has been identified by studies as present in both sauerkraut *and* sourdough starter. But of course, both sauerkraut and sourdough starter have dozens of other species besides! (See the chart on the next page for the ones I checked.)

Now, obviously you're not going to get live cultures from sourdough bread once it's baked. But if you ever eat naturally fermented sauerkraut, pickles, or the like, then a full-spectrum probiotic yogurt will be mostly redundant. You will likely be ingesting bacteria species already resident in your gut. (And yes, that includes those two bacteria du jour, L. reuteri and L. gasseri. They've both been identified in naturally fermented sauerkraut as well as in sourdough starter, as also shown in the chart.)

Wait, there's more! Though fermented foods are a concentrated source of lactic acid bacteria, you'll get *some* amount just from eating fruits and vegetables in their native state—that is, raw, whole, not heavily exposed to pesticides or disinfectant. For that matter, you might get them from contact with dirt!

But at least for the sake of discussion, let's say you've had limited sources of lactic acid bacteria, or you need to reacquire them. Then let's look at how you could optimize yogurt as a probiotic.

Cross-Culture Bacteria

	COMMERCIAL YOGURT	NATURAL SAUERKRAUT	SOURDOUGH STARTER
S. thermophilus	✓	✓	✓
L. bulgaricus	✓	✓	✓
L. acidophilus	✓	✓	✓
B. bifidum/L. bifidus[1]	✓	✓	✓
B. lactis/B. animalis[2]	✓	✓	✓
L. casei	✓	✓	✓
L. paracasei	✓	✓	✓
L. rhamnosus	✓	✓	✓
L. reuteri	—	✓	✓
L. gasseri	—	✓	✓

✓ = May be present   — = Not commonly found

1. L. bifidus is an older name for the species now classified as B. bifidum.
2. Both names are common older identifiers for what are now technically classified as two subspecies of B. animalis.

What It *Really* Takes to Make Yogurt

You may have heard the saying, "It takes yogurt to make yogurt." In other words, to make yogurt, you must have some yogurt on hand to start it with.

That's the common wisdom. So, for instance, if you want to make naturally fermented yogurt with a wider range of bacteria species, you're told you need one of the European "heirloom" yogurts. That means buying a packet of freeze-dried culture, or else finding someone to share their starter.

But luckily, the old saying is not true! You *don't* need yogurt to make yogurt. All you really need is a good natural source of lactic acid bacteria. With that, you can start your yogurt from scratch, right there in your kitchen. And it will include many more bacteria species than any yogurt you can buy in a store.

I've started yogurt from scratch with not one but several methods. I've also developed a hybrid approach that may give the best results of all. And once you know what I've done, I'm sure you'll have many ideas of your own.

Be aware, though, that yogurt you start from scratch may sometimes—but not always—be thinner than you're used to making. That's because, unlike with conventional yogurt, you're not getting a set of bacteria species chosen largely for how well they firm up the yogurt.

It's like the difference between a carefully cultivated garden and a field of wildflowers. Each has its own value, benefits, and delights, and you can't really say one is "better" than the other. In your own diet, you may well want both!

What About Raw Milk?

Anyone with broad experience in fermenting might tend to think of letting a food's natural microbes do all the work. Which might well lead to the question, why not make a full-spectrum probiotic yogurt by simply incubating raw milk? Since it hasn't been pasteurized, shouldn't this milk contain all the microbes needed to turn milk into yogurt?

The idea is reasonable enough that I had to try it, but I found it didn't work that way—as I would have known if I'd also had experience in cheese making! Raw milk certainly contains microbes that make yogurt, but it also contains competing ones that work against them, plus active enzymes that have their own effects on proteins.

If you incubate raw milk, or even just leave it at room temperature, then yes, it will turn sour and thicken. Some generous souls have termed this *yogurt*, but the correct term is *clabber*—according to Sandor Ellix Katz in his masterful book *The Art of Fermentation*.

In any case, it likely won't firm up much at first, and if you keep it going too long, the curds and whey may then separate dramatically, giving you a simple cheese. (That's what happened to mine.)

If you look at recipes for raw milk yogurt, they will likely tell you to enjoy your "runny yogurt." Or advise you to add a thickener. Or tell you to heat the milk high enough to kill off the native microbes and disable the enzymes before adding commercial yogurt or other starter—in other words, to first pasteurize it yourself.

That's not to say that fermented raw milk isn't tasty and healthful. It is, after all, the original source of countless dairy products, including most cheeses. But one thing it *can't* give you is yogurt!

Starting with Plants

My first method takes advantage of the fact that *any* plant grown in the open is naturally coated with a broad array of lactic acid bacteria ready to go to work. So, a wide variety of plants can be a source of bacteria for your yogurt. All you have to do is get them in the right form and give them the right conditions.

You do need to pay attention to that "right form," though. Any produce that's been cooked or had an outer layer removed is useless. The same with any plant processed with high heat or strong chemicals. You should also avoid produce coated in wax. Pesticides are less of a problem today than they once were—but still, if the plant is "organically" grown without them, so much the better.

My general method for starting from a plant source is much the same as my standard way of making yogurt, with just a few differences:

1. Instead of yogurt starter, add half a cup of your plant source to your milk. If needed, first cut the plant into manageable pieces. Do *not* skin or peel the plant or wash it with anything but cool to warm water.

2. When your culture reaches incubation temperature, strain out the plant source, then continue warming. In my proofer, reaching this temperature usually takes about three hours, but it may be less for you. If the time is *too* short, you may have to let the plant soak a little longer.

3. Incubation takes a little longer than usual. For me, it's the time to reach incubation temperature *plus* about another nine hours.

Sound familiar? Despite the different goal, this is basically a method of infusion, as described in my chapter on adding flavors. In this case, though, you leave out the yogurt starter entirely and make sure your added ingredient has live microbes on it.

In case you're tempted to just leave the plant source in the milk to become part of the yogurt—be careful! With fruit especially, the plant's natural sugar can encourage yeast and tilt the ferment toward the alcoholic. Also, if any pieces of a plant float to the top of the milk, they can get moldy. Removing the plant early is the safe way to avoid any problem.

Of course, not all potential plant sources will reach you in a form that works for you, and some will be easier to handle than others. So, I'll discuss some I've had success with, and some failures. All were used with the general method described above.

Some of these sources impart a distinctive flavor, while others have little taste of their own. Keep in mind, though, that by the time you use any yogurt as starter for a second generation, the taste will be diluted enough that you're not likely to notice it.

And speaking of that second generation: My experiments with these plant sources were mainly to provide "proof of concept." That means my goal was just to show they worked! Though there's no reason to think these yogurts could not start later batches, I haven't tested that myself.

Raisins

My very first experiment was with sun-dried, unsulphured raisins. These, I figured, provide a cheap, concentrated, convenient, year-round reliable source of lactic acid bacteria. Sure enough, they worked beautifully to firm the yogurt. And the taste they imparted is one you're not likely to mind! (I used Sun-Maid California Sun-Dried Raisins.)

By the way, don't throw away the raisins you remove. Milk-soaked raisins are delicious. Hot tip: Try them mixed back into your finished yogurt!

Though raisins are common and inexpensive, you can no doubt find other dried fruit that works just as well.

Carrots

One of my final experiments turned out to be one of my most successful. I took a raw, unpeeled, gently washed carrot and just sliced off enough pieces to fill one cup. (I used more than my usual half cup because of the big air spaces between slices.) The yogurt firmed nicely and had a mild, pleasant carrot flavor for this first batch.

Chickpeas

Dried beans are a good source of lactic acid bacteria, because no part of the bean has been removed. They're also easy to handle, and they impart little taste of their own. Remember, you *want* what's on the beans, so there's no need to rinse them if they already seem clean.

I had great success with chickpeas (aka garbanzo beans), which my wife happened to have on hand. Neither of us could detect any hint of bean flavor.

Lentils

You would think, if beans worked, lentils would too. But though they firmed the yogurt nicely, they left an unpleasant taste.

Cashews/Peanuts/Nuts

With their shells removed, nuts and their close relatives are *not* a good source of bacteria—because the bacteria are on the shells! I had zero success, for instance, with raw, shelled peanuts.

But cashews are a different story, because they're not a true nut. Instead, they're fruits that grow *without* shells. So, raw cashews ferment easily, besides having a mild, pleasant flavor that works perfectly with yogurt.

After you strain out the cashews, give yourself a special treat: Rinse them off, then roast them yourself. I roasted mine in an air fryer, which took five minutes at 300°F (about 150°C) for the half cup of them that started the yogurt. Delicious!

Whole Grains

Readers of my book *Smart Sourdough* will know that whole grains always come with the lactic acid bacteria needed to start a sourdough culture. So, those grains should be able to start yogurt too, right? Well, of course, they can! But they wouldn't be my first choice, or my second one, either.

Using my standard amount of half a cup as starter, I tried wheat berries, rye berries, and long grain brown rice. The wheat and rye took about three hours longer than the plant sources in my other experiments, and though the protein mesh formed, it seemed thin and soft. The brown rice took only an hour extra and was a little firmer. But all three yogurts tasted bad, and I wound up discarding them.

Vanilla Beans

Believe it or not, you can start a quart jar of yogurt with a single vanilla bean—and infuse its flavor at the same time! For a stronger flavor, of course, you can use more than one. Vanilla beans aren't cheap, but you should be able to use each of them several times over.

Ginger Root

I thought this would be a slam dunk. But, due to an enzyme in raw ginger root, a half cup of thin slices curdled three cups of milk before they even got up to incubating temperature.

Starting with Brine

As we've seen, adding lactic acid bacteria directly from a plant source is one way to get a wide range into your yogurt. Another way is to add a diverse bacteria culture that's already fully active. And one simple way to do that is to add a little brine from another natural ferment. That could be a ferment you buy, or one you make at home.

Since the bacteria are more concentrated in brine than on plant sources, you don't need to add much. That means you probably won't detect the added taste—and almost surely not in later batches started with that yogurt. (Of course, you might have more trouble if your brine is strongly seasoned.)

When does it make sense to use brine? If you already have it on hand, then adding it is simpler than starting from a fresh plant source. In fact, it's simple enough that you might prefer it for starting later batches too, in place of yogurt from the batch before.

My general method for starting from brine is again much the same as my standard way of making yogurt, with just a couple of differences:

1. Instead of yogurt starter, add one tablespoon of brine to your milk.

2. With this amount, incubation can take about three hours longer than with a yogurt starter. In my proofer, that's about twelve hours total.

Beyond the usual suspects of cabbage for sauerkraut and cucumbers for pickles, many other raw plants or parts of them can be fermented into usable brines—but not all. Fruits, for instance, have a high sugar content that encourages yeast, turning those ferments alcoholic. Brine with too much alcohol or too many competing microbes will curdle your milk, separating it into curds and whey instead of turning it into yogurt. The enzymes in seeds, activated and released with moisture, might also curdle milk. Sometimes, you might just need to experiment.

Luckily, to make your own ferments and brine, you don't need to follow the traditional days-long, cool-temperature methods. Instead, you can ferment at the same temperature you use for yogurt, with much the same equipment, in just 24 hours. Though each plant will have its quirks, there are just two essential rules to follow overall, to discourage competing microbes:

1. Keep the plant entirely submerged in liquid with a fermentation weight—bought, made, or improvised. Most fermentation weights are made specially for wide mouth Mason jars. For added water, you can use plain tap water, or filter it, or just let it stand a day or two to let the chlorine evaporate.

2. Add salt. Any kind will do, including iodized. For sauerkraut and other tamped-down ferments, I recommend one teaspoon per pound of produce. For pickles and other water-bath ferments, I recommend one tablespoon for a full quart Mason jar.

In my regular yogurt making, I doubt I'll ditch commercial yogurt as a starter and use brine instead. But after my experiments, I couldn't help thinking, *Why not?*

Sauerkraut

Sauerkraut brine works just fine for starting yogurt—and that's lucky, since if you're fermenting *any* vegetable at home, it's likely to be sauerkraut. It's also the vegetable ferment you're most likely to find in a grocery store, as it becomes more and more available in refrigerated deli sections.

Homemade sauerkraut

Bubbies Sauerkraut

Please understand, though, you can only use brine from sauerkraut that's *naturally fermented*. That means no vinegar! And store-bought sauerkraut with no vinegar can *only* be offered in refrigerated sections. (The same warnings apply to pickles.)

If you're buying sauerkraut in the United States or Canada, I can recommend an old favorite: Bubbies. Though the company purposely weakens the live cultures for long-distance transport, it took me no longer to make yogurt from Bubbies than from the brine of my own homemade sauerkraut. What's more, their cabbage is so mild, I couldn't tell from the yogurt's taste that I'd used brine at all.

Cashews

A simple homemade brine from raw, whole cashews gave me some of the best yogurt I've ever made—but not consistently! Brine from two later attempts—and even from the first one, after sitting awhile in the refrigerator—curdled the milk instead of making yogurt.

What happened? I can't say for sure, but I suspect at least part of the problem was enzymes. Cashews, after all, are seeds, so making the brine may have activated enzymes that ruined the yogurt. The first brine, when it worked, was fresh that day, so the enzymes had less time to develop.

With the yogurt from *that* brine, I had no trouble starting a second batch. This one came out as well as the first, and it took no longer to ferment than if I'd used a regular yogurt starter.

One benefit of this experiment was that it taught me the simplicity of making small ferments in a wide mouth *pint* Mason jar. A standard fermentation weight closely fits not only the mouth, but also the entire interior. That meant I had less worry about cashew pieces getting past the weight and floating to the top.

Pint Mason jar and weight

Cashews fermenting

Starting with Kefir Grains

Kefir is a dairy product that's soured and thickened but still drinkable. In its traditional and most diversely probiotic form, it's fermented at room temperature with *kefir grains*—soft clumps of living bacteria, yeast, and other fungi—that grow and multiply with use. The yeast give kefir a mild carbonation and a touch of alcohol that you don't get with yogurt.

After reading about kefir grains in David Asher's enlightening book *The Art of Natural Cheesemaking*, I wondered if they could be used to start yogurt as well, just by incubating at a higher temperature. From my sourdough experiments, I knew that yeast activity drops off as you reach the temperatures where the most heat-loving lactic acid bacteria start to thrive. So, maybe I could encourage the yogurt-making bacteria in the grains while discouraging the yeast.

For my experiment, I purchased kefir grains online from a most congenial U.S. company, Fusion Teas. I treated the grains just like my plant sources before, adding them to the milk at the start, heating to my desired temperature, then straining them out before continuing to incubate. (Note that I completely ignored the advice I got from the company and everyone else to "reactivate" the grains before use.)

Sure enough, at the end of another twelve hours, I had firm yogurt. After cooling, though, I detected a faint but unwelcome taste of alcohol—possibly from yeast activity as the yogurt cooled. That was enough for me to abandon the experiment, leaving further exploring to those less alcohol-averse. (Would it reduce the alcohol, for instance, to cool the yogurt more quickly by putting it in the freezer?)

Dual Culturing

So far, we've looked at ways to start a full-spectrum probiotic yogurt from scratch—in other words, to add bacteria where nearly none were before.

You need such an approach, of course, for experimenting. But in everyday practice, there's no need to be a purist. After all, our goal here is to *add* to the number of bacteria species, not *remove* any. So, why not start with bacteria found in commercial yogurt, then just supplement them?

Any alternative starter recommended in this chapter can also be used in dual culturing. Just add your plant source or brine to the milk along with your regular yogurt starter. You won't need as much of the plant or brine as if used alone—say, a quarter cup instead of a half for a plant, or a teaspoon instead of a tablespoon for a brine. Then, if you added something solid, strain it out once the culture reaches your normal incubating temperature, before continuing normally.

With two starters, incubation time should be slightly shorter than the time needed with your regular starter alone—but maybe not enough to notice. Remember, incubation time depends less on starter amount than on fermentation speed, which is based on the bacteria's doubling time.

Though I don't recommend focusing on a specific bacteria species, you could also use dual culturing to add one of those. Making good yogurt from a single species like L. reuteri might be a challenge, but it's no trouble at all to empty a capsule of it into your usual yogurt culture. That way, you get good yogurt *and* L. reuteri—for what it's worth.

Good Practices

In this chapter, I've purposely left you a lot of room for experimenting. If you go that route, I suggest you take advantage of a couple of standard scientific practices, to avoid the kind of false conclusions so common among those who try new kinds of yogurt.

The first practice is to either sanitize or sterilize your equipment to remove microbes that might influence your result. *Sterilizing* means destroying every single microbe, while *sanitizing* means reducing their number to a point where they won't cause any trouble—and this is usually good enough.

You can sanitize, for example, by soaking your equipment in hot tap water, if the temperature hasn't been lowered to avoid scalding; or by wiping it with 70% isopropyl alcohol. You could also use a dishwasher's sanitize cycle, or maybe even just its high-temperature setting.

The other practice is to use a *control*—a parallel experiment with everything the same *except* the one thing you're testing. Say, for instance, you're testing to see if a dose of brine will firm up your milk—which, because of the brine's weakness, winds up taking over a day. In that case, you want to *also* make sure the milk wouldn't firm up in the same time by itself. So, alongside that test, you incubate the same amount of milk *without* the brine. That's the control.

. . . and Bad Ones

I was reminded of the importance of these practices—and especially of the need for a control—when I explored the "yogurt" recipes of Dr. William Davis, as given in his book *Super Gut*.

These recipes are meant to produce yogurt with high probiotic value. But from my very first reading, I was convinced that Dr. Davis hadn't a clue about cultivating bacteria.

Davis's method is based on the notion that the population growth of bacteria proceeds in a straight line—in other words, the longer you let it go, the more bacteria you get. But in fact, that growth occurs in a cycle. Yes, the population at first increases rapidly. But eventually, the population gets too big to support itself on the available food or succumbs to its own toxic waste—the same lactic acid we want it to produce!

At that point, the population starts to decline and die off. In a single-species culture warmed for 36 hours—the fermenting time Davis recommends—you might not have many bacteria left alive.

It's true that Davis reported high bacteria counts from lab tests, but this does not really help his case. As a little research revealed, his lab's counting technology—called *flow cytometry*—does not normally distinguish between living and dead bacteria. And notably, his book never says he counted only the living.

Davis seems more up front about this in an online article, "The Arithmetic of Yogurt." While presenting calculations of L. reuteri population growth, he tells us that, for the sake of simplicity, "We are not accounting for bacterial death." He then goes on to assure us that even dead L. reuteri is good L. reuteri!

Davis himself warns us that first batches of his recipes are likely to fail, separating into curds and whey. That didn't surprise me, considering his extra-long incubation and his use of half and half in place of milk. To me, that sounded a lot like a way to make a simple cheese.

What puzzled me, though, was that Davis's followers reported success with starting later batches from their first one, when the original probiotic *should* have been dead or severely weakened. According to them, what they were getting was extremely firm and a bit tart. How was that possible?

I finally concluded that the key to the firmness, if not to the taste, was the inulin—an ingredient included in every Davis recipe, but one I had never used or researched. Davis says the inulin is added to provide food for the bacteria during the long ferment. But in commercial yogurt, I learned, it's used as a stabilizer like pectin or gelatin. It can hold yogurt together if the protein mesh starts to break down, or even solidify a culture that *never* formed the mesh.

In the end, it looked like Davis's concoctions weren't yogurt at all. They were inulin pudding!

It's one thing to know something for yourself, and another to be able to convince others. For that, I knew I had to do what Davis and his fans seemed to studiously avoid: set up a control. So, I made Davis's L. reuteri yogurt recipe, using inulin and half and half, just as he instructed—but I left out the L. reuteri.

The result? What I got smelled sour, tasted tart, and as the photo shows, came out completely firm.

L. reuteri "yogurt"—but with no L. reuteri!

To me, this was proof it was inulin, not fermentation, that was firming up Davis's yogurts. But for members of one Facebook group for probiotic yogurts, this photo was not convincing. Most commenters insisted my culture had become contaminated by other bacteria—from my utensils, or the air in my kitchen, or possibly from the half and half itself.

I had to admit I'd been lax with sanitation in this particular experiment. But I was also intrigued by the idea of contamination from my ingredients—and not just the half and half. Since inulin is a plant product, I wondered if any lactic acid bacteria might be sneaking in on it.

I wrote to the manufacturer of my inulin to ask about that. They assured me their product was sterile, due to the heat used to process it. So, no complications there.

Further research on dairy "contamination," though, surprised me. I hadn't realized that even pasteurized milk and cream carry enough live bacteria to eventually sour and thicken them. I learned that cream will turn into crème fraîche, and milk will turn into *clabber*—not really yogurt, but something you might mistake for it—while combined milk and cream, as in half and half, will turn into sour cream.

After 36 hours, then, there might well be *some* kind of fermentation, just from the recipe's half and half. And that fermentation, instead of the inulin or any L. reuteri, would be the likeliest source of this recipe's tartness.

To test this, I now needed a control for my control. In other words, after experimenting with just half and half plus inulin, I now needed to experiment with half and half by itself! (And this time, with careful sanitation.)

Just like my culture with inulin, the half and half alone started firming in 12 to 15 hours. And in the end, it tasted about the same—not like any yogurt I've tasted, but definitely tart. The

L. reuteri "yogurt," but with no L. reuteri or inulin

difference in the final products, though, was that the half and half with inulin had shown very little whey separation, while the half and half alone—again, as seen in the photo—showed a lot!

By now, it seemed clear it was inulin giving Dr. Davis's "yogurt" its dramatic firmness, while remaining bacteria in the half and half were enough to make it tart. So, neither of these qualities gave any guarantee of L. reuteri's presence, while the recipe could hardly do better at masking its absence.

Full disclosure: I never did make Davis's recipe with any L. reuteri!

Strength or Diversity?

If you were expecting this chapter to provide special recipes for cultivating single species of lactic acid bacteria, I'm sorry to disappoint you. The truth is, to cultivate any such species—L. reuteri, L. gasseri, and the like—you shouldn't need a special recipe at all. Just try one of my recipes from earlier chapters and replace the usual yogurt starter with your purchased powder.

Yes, for L. reuteri, you should probably lower the temperature a bit—maybe to around 100°F (38°C)—and you might have to wait a few hours longer for setting, if the starter culture is weak. But the method would be basically the same.

But is a single-species culture even what you want?

The recipes in this chapter should produce cultures that include L. reuteri, L. gasseri, and a whole lot more. And the whole point of the microbiome is that it's a wide variety of microbes *cooperatively* providing benefits for your body. Flooding the microbiome with a single species—if that was even possible—could be counterproductive.

What about the many personal reports of health benefits from eating single-species cultures? These benefits, when real, are more likely due to the inulin usually included in these recipes. Inulin is a true miracle food with benefits to both body and mind.

But before you go adding inulin to my recipes and spoiling the taste, consider that inulin is likely to be simplest and cheapest to get from foods where it's found naturally. Absolutely must have it with your yogurt? Just slice in some banana!

Scientists have barely begun studying all the actions and interactions that take place in your gut. But the main take-away has been that good health is tied to a greater *diversity* of microbes—not to the dominance of one species or another. Trying to boost any single species is just another instance of the piecemeal approach to nutrition that has so utterly failed us over recent generations.

APPENDIX
The Non-Dairy Option

There's no way around it: The modern dairy industry is a hotbed of health concerns and ethical issues.

Whether you're talking about contamination by hormones, antibiotics, and pesticides, or the killing of unwanted calves, or the confinement and crowding of animals, or the contribution to deforestation and global warming, eating dairy can be a hard sell. It's no wonder, then, that many are looking for dairy-free, plant-based alternatives.

Of course, when they do, they want to bring along their favorite foods—and one food they love to adapt is yogurt. But contrary to claims, yogurt is *not* something you can make from most plant milks.

Can Yogurt Be Vegan?

True yogurt is made from proteins linking up in a firm mesh in reaction to acid produced by lactic acid bacteria. The main protein that does this is casein, a protein found only in animal milks.

Some members of the globulin family of proteins also react like this, especially if heated beforehand. (You may remember that scalding can activate the lactoglobulin in dairy milk, providing extra help with forming the protein mesh.) But most plants don't have this kind of globulin. In fact, among the plant milks common today, the only one that *does* have it is soy.

Even with the right protein, you face a second challenge: Solids in plant milk don't stay suspended in liquid as well as solids in dairy milk do. They like to settle out. With a homemade plant milk, this means the solids won't stay evenly distributed for as long as it takes yogurt to firm up.

The upshot is, few recipes for dairy-free "yogurt" have anything to do with forming a protein mesh. In place of that, they tell you to firm up the culture by adding a stabilizer or thickener, or by thickening the starch already in the plant—even when you're adding live cultures for souring.

So, what you really get is sour pudding that's a yogurt *substitute*. And the same goes for when you buy most kinds of dairy-free "yogurt" in the store.

There's nothing wrong with substitutes! But while researching this book, I wanted to know if a *true* yogurt could be made from plant milk—and I found that some people do just that.

Success with Soy

Here's the trick to making a true vegan yogurt: Use a commercial soy milk with no additives—in other words, store-bought soy milk that's just soy and water.

Most commercial soy milks—like other commercial plant milks—use stabilizers to keep solids suspended. In regard to health, these additives are generally harmless or even beneficial, but they can hinder the forming of a protein mesh. Luckily, several companies still make soy milk without such additives and instead just *homogenize* it—just like dairy milk! (That's something you likely can't do well at home.)

In the United States, three companies with "pure" soy milk are West Life, Edensoy, and Pacific Foods. Read the labels, though, as the same companies may also sell varieties with additives. (Generally, the "unsweetened" varieties are the pure ones.) The product I chose for my own experiment was West Life Organic Unsweetened Soymilk, Plain.

A "pure" soy milk

> # Smart Soy Milk Yogurt
>
> 3 cups unsweetened commercial soy milk with no additives
> 1 tablespoon plain yogurt with live cultures
>
> With these ingredients, follow the directions for Smart Yogurt.

I used the exact same recipe as for my dairy yogurt, just substituting soy milk for cow. Not being a purist, I also used my regular commercial yogurt as starter—though any vegan "yogurt" with live cultures should work instead. So should the go-to bacteria source of many vegan yogurt makers: a probiotic supplement with multiple species. Of course, whatever you choose, yogurt from the first batch should be able to start the next.

The yogurt from this recipe was pleasant enough, with a slight nutty flavor. Soy milk, though, is only around 2% fat—the same as low-fat milk—so, I found the taste and texture a bit thin.

If you're used to that consistency, you might well enjoy it—especially if you're adding other ingredients. For that matter, you could raise the fat content by stirring in some coconut milk before eating. I found that combination quite pleasant, after adding a little maple syrup.

An even simpler way to add fat: Mix in some vegetable oil, again before eating. A teaspoon of oil mixed into a cup of yogurt will add about 2% to the fat content percentage. I tried sunflower oil, and it definitely improved the taste and texture.

In place of the starters mentioned above, you should be able to use any of the alternative bacteria sources from my chapter on boosting probiotics, *all* of those being vegan. Sources like that will give you the greatest diversity of probiotics—much greater than you'll get from a tablet.

Index

For recipes, please see the table of contents. Recipe ingredients are indexed only when appearing in background discussions or recipe tips.

A2 milk, 70
A2 Milk Company, 70
acidity, 22, 64, 69. *See also* sourness *and* lactic acid
air fryer, 80
alcohol, 47, 50, 59, 78, 81, 85–86
"Arithmetic of Yogurt, The" (article), 88
Art of Fermentation, The (book), 76
Art of Natural Cheesemaking, The (book), 85
Asher, David, 85

B. animalis (Bifidobacterium animalis), 74
B. bifidum (Bifidobacterium bifidum), 74
B. lactis (Bifidobacterium lactis), 41–42, 74
bacteria. *See* lactic acid bacteria
basil, 60
beans, dried, 79
beet juice, 63
benefits, 5, 13–14, 17, 38, 65, 70, 92
beverage powders, 60–62
blender, 18, 51
blue tea, 64
brine and brines, 81–84, 86–87

Brød & Taylor Folding Proofer & Slow Cooker, 30–31, 33
Bubbies Sauerkraut, 83
Bulk Supplements, 40
butterfly pea flowers, 64

cacao nibs, 59, 61
calcium, 39
Cambro Manufacturing, 21
Canada, 35, 83
caraway seeds, 59
Carlisle Food Service Products, 21
carrots and carrot juice, 63, 79
casein, 7, 36, 40, 70, 93. *See also* milk protein and proteins
casein powder, 40
cashews, 57, 79–80, 83–84
cheese and cheese making, 7, 76, 88
chickpeas, 79
Chobani yogurt, 36, 41
chocolate milk, 61
citric acid, 55
citrus fruits, 55, 60, 63
clabber, 76, 90
cocoa, 57, 61–62, 64
coconut milk, 95
coffee, 57, 60, 62, 64
colors and coloring, 48, 63–64
condensed milk, sweetened, 37

contamination (by microbes), 22, 90
control, experimental, 87, 89–90
coriander seeds, 60
cornstarch, 40
cream, 43–44, 48, 90
crème fraîche, 90
crocks, 2.7-quart, 21, 28, 30
cumin seeds, 59–60
curdling, 47, 55, 59–60, 63, 80–82
curds, 76, 82, 88
cytometry, flow, 88

dairies and dairy industry, 43, 93
dairy fat. *See* milk fat
dairy-free yogurt, 6, 13, 40, 69, 93–95
dairy intolerance, 5–6, 38, 49, 65–71
Davis, Dr. William, 40, 87–91
denaturing, protein, 7, 40, 93
dietary fiber. *See* fiber, dietary
dieting and weight consciousness, 13–14, 43–44, 48–50
digital oven, 32
doubling time (of bacteria), 9–10, 20, 86
dressings, 48–49, 57
dried beans, 79
dried fruit, 79
dry milk powder. *See* milk powder, dry

Edensoy soy milk, 94
electric yogurt makers. *See* yogurt makers, electric
enzymes. *See* milk enzymes *and* plant enzymes
espresso, 61, 64. *See also* coffee
Europe, 35, 75

evaporated milk, 37
extracts, flavor, 47, 50
Fage yogurt, 36, 41
Fairlife milk, 38, 56–57, 63–64, 67
fat. *See* milk fat
fermentation weight, 82, 84
fermented foods, 8, 11–12, 33, 47, 72–73, 76
fermenting and fermentation, 7, 9–12, 16–19, 21, 28–29, 33, 47, 49, 57–58, 66–69, 73, 76, 82–84, 86, 88–90
fiber, dietary, 41, 71, 73
firmness, 6–7, 9–10, 12–13, 19, 21, 35–46, 88–91
flavor extracts, 47, 50
flavors and flavoring, 6, 16, 38–39, 47–64
flow cytometry, 88
FODMAPs (fermentable oligosaccharides, disaccharides, monosaccharides, and polyols), 71
food safety. *See* sanitation *and* spoilage and preservation
French Women Don't Get Fat (book), 44
French yogurt, 8, 41, 45
frother, 18, 51
fruit, 47, 79
fruit, dried, 79
Fusion Teas, 85

GAPS Diet (Gut and Psychology Syndrome Diet), 69
garbanzo beans. *See* chickpeas
gelatin, 18, 40, 89
Ghirardelli Double Chocolate Hot Cocoa Mix, 61

ginger and ginger root, 59–60, 80
globulin, 93. *See also* lactoglobulin *and* plant protein and proteins
glycerin (or glycerine), 50, 52
glycerol, 50
glycol, 50
goat milk, 43
Golden Milk, 64
grains, whole, 80
grape juice, 63
Greece, 8, 36, 43
Greek yogurt, 8, 36, 41–43, 45, 56, 63, 72
ground vanilla. *See* vanilla powder
Guernsey cows, 43, 70
Guiliano, Mireille, 44
gut health. *See* microbiome and gut health *and* probiotics

half and half, 43, 45, 48, 61, 88–91
heat-loving bacteria. *See* thermophilic bacteria
heirloom yogurt, 9, 12, 75
Hershey's 100% Cacao Natural Unsweetened Cocoa, 62
Hershey's 100% Cacao Special Dark Cocoa, 62
Holstein cows, 43
homogenization, 15, 43, 94

IBS (irritable bowel syndrome), 71
incubating and incubation, 5, 9–10, 12, 16–34, 47, 58, 68–69, 77, 81, 85–86, 88, 92
infusion and infusions, 57–60, 77
Instant Pot, 28–29, 31, 35

intolerance, dairy. *See* dairy intolerance
inulin, 12, 40–41, 89–92
Jersey cows, 43, 70
juice and juices, 39, 53–57, 63–64
juice concentrate and concentrates, 54–56, 63
juice powder and powders, 54, 56, 60, 62–64

Katz, Sandor Ellix, 76
kefir and kefir grains, 85–86

L. acidophilus (Lactobacillus acidophilus), 41–42, 74
L. bifidus (Lactobacillus bifidus), 74
L. bulgaricus (Lactobacillus bulgaricus), 41–42, 74
L. casei (Lactobacillus casei), 41–42, 74
L. gasseri (Lactobacillus gasseri), 73–74, 92
L. paracasei (Lactobacillus paracasei), 74
L. reuteri (Lactobacillus reuteri), 33, 73–74, 86, 88–92
L. rhamnosus (Lactobacillus rhamnosus), 41–42, 74
Lactaid milk, 67
lactase, 39, 65–68
lactic acid, 7, 11–13, 36, 68–69, 88, 93
lactic acid bacteria, 6–13, 17–18, 20, 22, 33, 36, 41–42, 45, 47, 49, 66–69, 72–81, 85–95
lactoglobulin, 36, 93

lactose, 5, 11, 38–39, 48–50, 65–71. See also milk sugar and sugars and lactose powder
lactose-free milk, 11, 38–39, 49, 65–67, 70
lactose-free yogurt, 5, 66–69
lactose intolerance, 38, 49, 65–69, 71. See also dairy intolerance
lactose powder, 38, 40, 49, 53, 56, 61
lemons and lemon juice, 55, 60, 63
lentils, 79
lime juice, 55
"live cultures," 5, 11, 15–16, 41–42, 83, 94–95
low-fat milk, 13–15, 35–36, 95
low-fat yogurt, 10, 13

maple syrup, 95
Mason jars, 16, 24–25, 28, 30, 32, 82, 84
McGee, Harold, 35
"Mediterranean" yogurt, 45–46
microbiome and gut health, 5–6, 11–12, 41, 65, 71–73, 92. See also probiotics
microwave oven, 51, 58
milk cartons, 25–27
milk concentrate, 37–39
milk enzymes, 35, 76
milk fat, 7, 13–15, 39, 43–45, 47, 56
milk powder, dry, 14, 37
milk protein and proteins, 7, 10, 13, 21, 35–40, 55–56, 63, 70, 76, 89, 93
milk sugar and sugars, 7, 11, 21, 38–39, 48–50, 65–69, 71
mold. See spoilage and preservation

New York Times, The, 35
non-dairy yogurt. See dairy-free yogurt
nonfat dry milk powder, 37. See also milk powder, dry
nonfat milk, 13–15, 35–37, 39, 43
nonfat yogurt, 10, 13–14, 35, 43
North America, 35. See also United States *and* Canada
Now Foods, 38, 49
Nutricost Lactase Powder, 68
nutrition, 5, 7, 9, 13, 21, 36, 38–40, 67, 92
nuts, 79

On Food and Cooking (book), 35
orange juice, 55, 63
oregano, 60

Pacific Foods soy milk, 94
pasteurization, 10, 15–18, 35–36, 43, 76, 90
peanuts, 57, 79
pectin, 18, 40–41, 89
peppermint, 59, 62
pickles, 11–12, 47, 73, 81–83. See also fermented foods
pineapple juice, 55
plant-based yogurt. See dairy-free yogurt
plant enzymes, 55, 82–83
plant milk and milks, 13, 40, 93–95
plant protein and proteins, 13, 40, 93–94
plant starch, 94
population growth (of bacteria), 9–10, 12, 88

Post-it tape, 18
powdered milk. *See* milk powder, dry
prebiotics, 41, 71, 73
preheating. *See* scalding
preservation. *See* spoilage and preservation
probiotics, 5–6, 11–12, 40, 69, 72–92, 95
proofer, home. *See* Brød & Taylor Folding Proofer & Slow Cooker
proteins. *See* milk protein and proteins *and* plant protein and proteins
protein intolerance, 70. *See also* dairy intolerance

raisins, 78–79
raw milk, 15, 76
raw milk yogurt, 76
rice, 80
rosemary, 60
rye berries, 80

S. thermophilus (Streptococcus thermophilus), 41–42, 74
salt, 82
sanitation, 10, 17–18, 87, 90. *See also* sterilization
saturated fat, 13
sauerkraut, 11–12, 22, 73–74, 81–83. *See also* fermented foods
scalding, 10, 14, 16, 18, 27, 35–37, 93
scaling (of recipes), 21
SCD (Specific Carbohydrate Diet), 69
sesame seeds, 59–60
sheep milk, 43

SIBO (small intestine bacterial overgrowth), 71
smart oven, 32
Smart Sourdough (book), 80
sour cream, 48, 90
sourdough and sourdough starter, 11, 73–74, 85. *See also* fermented foods
sourness, 7, 9, 13, 19, 21, 38, 40, 47–49, 68–69, 89, 94
sous vide and sous vide cookers, 25–29, 31, 33
soy milk, 93–95
spirulina powder, 64
spoilage and preservation, 16–17, 22
stabilizers. *See* thickeners and stabilizers
Starbucks Double Chocolate Hot Cocoa, 61
starter, yogurt, 9–11, 16 21, 33, 41, 45, 51, 55–56, 58, 67, 75–78, 81–83, 86, 88, 92, 95
sterilization, 15, 17, 87. *See also* sanitation
sterilized milk, 15
straining, yogurt, 6–8, 14, 28, 36–37, 41, 43
sugar (sucrose), 47, 49, 52. *See also* sweetening and sweeteners
sunflower oil, 95
Sun-Maid California Sun-Dried Raisins, 78
Super Gut (book), 87–88
supplements, probiotic, 12, 86, 92, 95
sweetening and sweeteners, 16, 38, 47–50, 53, 56, 68–69, 71, 95

Talcufon Vanilla Bean Powder, 52
taste, 5–6, 14, 35–46, 67, 78, 88–92.
  See also flavors and flavoring
tea, instant, 60
temperature, 9–12, 16, 18, 20–25,
  27–36, 51, 58, 85, 92
tests and testing, laboratory, 12, 69, 88
texture. See firmness and thickness
themes, 57, 60
thermometer, 5, 19, 51
thermophilic bacteria, 10–11, 33, 85.
  See also lactic acid bacteria
thickeners and stabilizers, 5–6, 12–13,
  18, 37, 40–41, 55, 76, 89, 94
thickness, 6–8, 14, 35–46, 55–56, 67,
  75, 94
tomato juice, 63
tomato paste or concentrate, 63
toppings, 48–49
turmeric root, 64

UHT milk (ultra-high-temperature
  milk), 15, 36
ultra-filtered milk, 15, 37–39, 56,
  63–64, 66–67
ultra-processed foods, 39
United Kingdom, 70
United States, 8–9, 35–36, 42, 66, 83
United States Food and Drug
  Administration, 66

Vanilla Bean Kings Madagascar
  Vanilla Beans, 53
vanilla beans, 50, 52–53, 57, 80
vanilla extract, 50, 52

vanilla flavoring, non-alcoholic, 50, 52
vanilla powder, 52, 61–62
vegan yogurt. See dairy-free yogurt
vegetable oil, 95
vinegar, 83

Watkins Organic Pure Vanilla Alcohol
  Free Flavoring, 50
Watson, Anne L. (wife), 9, 27, 37–38,
  57, 79
weight consciousness. See dieting and
  weight consciousness
West Life Organic Unsweetened
  Soymilk, 94
wheat belly, 71
wheat berries, 80
wheatgrass juice powder, 64
whey and whey separation, 7, 19,
  21–22, 36–37, 40, 63, 76, 82, 88, 91
whipped cream, 48
whisks and whisking, 18, 51
whole dry milk powder, 37. See also
  milk powder, dry
whole milk, 13–15, 35, 37–38, 56,
  63–64
wine, 47

yeast, 47, 78, 81
yogurt (defined), 7–9, 42
yogurt, commercial, 5–6, 8–9, 11–12,
  14–15, 18–19, 33, 35–37, 40–41, 56,
  72–76, 82, 86, 89, 95
yogurt maker, electric, 10, 24
yogurt, traditional, 8–9, 12, 14, 35, 43

zest, fruit, 60

About the Author

Mark Shepard is the author of two popular books on sourdough, *Simple Sourdough* and the revolutionary *Smart Sourdough*. He has also published books on nonviolent social change, simple living, and the flute, as well as children's books under the name Aaron Shepard.

Mark has made fermented foods—sourdough, yogurt, pickles, and more—for much of half a century. He now lives in Bellingham, Washington, with his wife and fellow author, Anne L. Watson. Visit him at

www.markshep.com

More by Mark . . .

ALSO FROM SHEPARD PUBLICATIONS

and more . . .

NOTES

www.ingramcontent.com/pod-product-compliance
Lightning Source LLC
Chambersburg PA
CBHW060927170426
43193CB00022B/2982